ARE AND
ED PLANTS OF
SAN MATEO AND SANTA CLARA COUNTY

with
Photographs and Illustrations

rolinon congestum

by

Toni Corelli and Zoe Chandik

THE RARE AND ENDANGERED PLANTS
OF SAN MATEO AND SANTA CLARA COUNTY
with
Photographs and Illustrations
Toni Corelli and Zoe Chandik

ISBN 0-9646994-0-0

Published by MONOCOT PRESS
Post Office Box 773
Half Moon Bay, CA 94019

TABLE OF CONTENTS

Map Showing San Mateo and Santa Clara County and Surrounding Counties

INTRODUCTION

Purpose of the Book:

We saw a need for a book of this type because our flora is increasingly threatened by urbanization, conversion of land to agricultural and industrial use, recreational activities and the invasion of non-native plants. Our counties, San Mateo and Santa Clara, include a diversity of habitats from coastal San Mateo County, part of the Santa Cruz Mountains, San Francisco Baylands and part of the Inner Coast Range. Along with this diversity of natural habitats is a variety of climatic conditions and elevation gradients. This may be why we have so many rare plants (106) in San Mateo and Santa Clara County. This is almost 10% of the total plants in the State CNPS Inventory. Most of the information in this book is available through the California Native Plant Society's (CNPS) Inventory of Rare and Endangered Vascular Plants of California, fifth edition. What we thought would be useful is the description of the plant, plant associate information and the photographs of the plant. We hope this book will be useful to Environmental Consultants, City and County Planners, Public Land Managers, the interested public, and others documenting the rare plants in our area.

The prime purpose of this publication is to acquaint the public with our local rare and endangered flora. In doing so we hope to increase public awareness of our botanical treasure and the need to protect it. This can lead to more effective protection and preservation for future generations to enjoy. We also became aware of the need to continue the process of documentation of the sensitive species. Some plants on this list may be more common than thought while others deserve greater protection by the State and Federal Government. A plant lost to extinction is lost forever. Some of our rare plants are on the brink of extinction because of human activity. It is essential to establish a compromise between conservation and development that will preserve these valuable entities before it is too late.

This will be the first photographic book of Rare and Endangered Plants by county that has been published in the State so far. Our local chapter of CNPS, the Santa Clara Valley Chapter has been very involved in the preservation of sensitive plants and habitats and this publication is a result of that activity.

Toni Corelli and Zoe Chandik

ACKNOWLEDGEMENTS

It has taken four years to accumulate the photographs and illustrations for this publication. We thank all of the people who have contributed to this endeavor. Much appreciation is accorded to the following people:

Photographs by:

Ted Chandik: Dirca occidentalis (close-up). **Zoe Chandik**: Acanthomintha duttonii (habitat), Acanthomintha lanceolata (close-up), Allium sharsmithae, Arctostaphylos andersonii, Balsamorhiza macrolepis macrolepis, Calochortus umbellatus (habitat), Calyptridium parryi hesseae, Campanula exigua (habitat), Ceanothus ferrisae (close-up), Cirsium fontinale campylon (close-up), Cirsium fontinale fontinale, Collinsia multicolor (close-up), Cupressus abramsiana, Dudleya setchellii, Eriogonum argillosum, Eriogonum luteolum caninum, Eriophyllum jepsonii (habitat), Erysimum franciscanum (close-up), Fritillaria agrestis, Fritillaria biflora ineziana (close-up), Fritillaria falcata (close-up), Galium andrewsii gatense, Helianthella castanea, Hesperolinon congestum (close-up), Lessingia micradenia glabrata, Linanthus ambiguus, Malacothamnus hallii, Pentachaeta bellidiflora (close-up), Psilocarphus brevissimus multiflorus, Streptanthus albidus peramoenus (close-up). **Toni Corelli**: Acanthomintha duttonii (close-up), Acanthomintha lanceolata (habitat), Ceanothus ferrisae (habitat), Chorizanthe cuspidata cuspidata, Cirsium andrewsii, Cirsium fontinale campylon (whole plant), Elymus californicus (close-up), Eriophyllum latilobum (habitat), Grindelia hirsutula maritima, Hesperolinon congestum (habitat), Lessingia arachnoidea, Lessingia germanorum, Lessingia hololeuca, Pedicularis dudleyi (habitat), Streptanthus albidus albidus. **B. Delgado**-BLM Hollister: Hemizonia parryi congdonii. **Barbara Ertter**: Horkelia cuneata sericea. **John Game**: Cordylanthus maritimus palustris, Linanthus acicularis. **Gloria Heller**: Arctostaphylos regismontana, Castilleja affinis neglecta (habitat), Clarkia concinna automixa, Pedicularis dudleyi (close-up). **Ken Himes**: Arabis blepharophylla, Arctostaphylos imbricata, Arctostaphylos montaraensis, Clarkia brewer (close-up), Collinsia multicolor (habitat), Cypripedium montanum, Dirca occidentalis (habitat), Eriophyllum jepsonii (close-up), Eriophyllum latilobum (close-up), Erysimum franciscanum (habitat), Fritillaria liliacea, Limnanthes douglasii sulphurea (close-up), Lupinus eximius, Malacothamnus arcuatus, Pentachaeta bellidiflora (habitat), Pinus radiata, Streptanthus albidus peramoenus (habitat), Streptanthus callistus (habitat). **Glenn Keator**: Delphinium californicum interius. **Don Mason**: Campanula exigua (close-up), Campanula sharsmithiae, Clarkia breweri (habitat), Coreopsis hamiltonii, Phacelia phacelioides,

Plagiobothrys chorisianus chorisianus, Plagiobothrys myosotoides, Sanicula saxatilis (habitat), Silene verecunda verecunda, Streptanthus callistus (close-up). **R. Morgan**: Perideridia gairdneri gairdneri (close-up), Piperia candida, Piperia michaelii (close-up). **Bart O'Brien**: Elymus californicus (habitat), Fritillaria biflora ineziana (habitat), Fritillaria falcata (habitat), Sanicula saxatilis (close-up). **Brad Olson**: Grindelia stricta angustifolia, Monardella villosa globosa, Plagiobothrys glaber (herbarium specimen), Senecio aphanactis (herbarium specimen), Trifolium amoenum, Tropidocarpum capparideum (herbarium specimen). **M. Perkins**: Piperia michaelii (habitat). **Mark W. Skinner**/California Native Plant Society: Castilleja affinis neglecta (close-up), Lasthenia conjugens (habitat), Legenere limosa (habitat), Lilium maritimum, Suaeda californica. **Doreen Smith**: Linanthus grandiflorus. **Dean Taylor**: Astragalus tener tener. **Mike Vasey**: Potentilla hickmanii. **Bob Weller**: Calochortus umbellatus (close-up). **California Native Plant Society Slide Library**: T.I. Chuang: Perideridia gairdneri gairdneri (habitat); Juanita Doran: Legenere limosa (close-up), Ranunculus lobbii (close-up); Wilma Follette: Triphysaria floribunda; M. Fruhling: Eriastrum brandegeae; D. Goforth: Cypripedium fasciculatum; James R. Griffin: Erysimum ammophilum (habitat); N.L. Mason: Lathyrus jepsonii jepsonii; Harry McAndrew: Ranunculus lobbii (habitat); Jake Ruygt: Lasthenia conjugens (close-up); R. Schlising: Limnanthes douglasii sulphurea (habitat); Rick York: Erysimum ammophilum (close-up), Horkelia marinensis.

Illustrations:

ILLUSTRATED FLORA OF THE PACIFIC STATES, Four Volumes, By LeRoy Abrams and Roxana Stinchfield Ferris. Used with the permission of the publishers, Stanford University Press. © 1960 by the Board of Trustees of the Leland Stanford Junior University: Atriplex joaquiniana, Calandrinia breweri, Chorizanthe robusta robusta, Equisetum palustre, Eriastrum brandegeae, Plagiobothrys glaber, Plagiobothrys uncinatus, Senecio aphanactis, Sidalcea hickmanii, Sidalcea malachroides, Tropidocarpum capparideum.

A FLORA OF THE MARSHES OF CALIFORNIA, By Herbert L. Mason. University of California Press, Berkeley and Los Angeles. © 1957 by the Regents of the University of California: Azolla mexicana, Potamogeton filiformis.

Other illustrations by:
Jennifer Fairfax: Sanicula hoffmannii.

Linda Miller: Hesperolinon congestum (title page).

Rare and Endangered Plants of San Mateo and Santa Clara County by Family

Apiaceae (carrot) ***Eryngium aristulatum hooveri***
Hoover's button celery
CNPS List: 4 **State/Federal Status:** /C1
Distribution: SBT, SCL, SLO
Habitat: Vernal Pools.
Description: Perennial herb, stout, ascending to erect, 1-9 dm tall, generally branching loosely from main stem. Leaf blade 3-10 cm long, 1-2 cm broad irregularly cut, or lobed. Heads in cymes, 5-23 mm in diameter; peduncles 0.5-1.5 cm long; bracts 5-8, 6-27 mm long, becoming longer in fruit; bractlets sharply toothed, exceeding flowers. Petals white; styles equal to calyx, sometimes purplish. Flowering time: May-August.
Plant Associates:
Comments: See *Madrono* 30(2):93-101 (1983) for original description. J147

Apiaceae (carrot) ***Perideridia gairdneri gairdneri***
Gairdner's yampah
CNPS List: 4 **State/Federal Status:** /C2
Distribution: DNT,HUM,KRN,LAS,LAX*,MEN,MNT,MOD,MRN,NAP, ORA*,SBT,SCL,SCR,SDG*,SIS,SLO,SMT(*?),SOL,SON,TRI
Habitat: Moist soil of flats, meadows, stream sides, grasslands and pine forests.
Description: Perennial up to 14 dm tall. Root tuberous, single, 10-15 mm wide. Basal leaf petiole up to 15 cm long, generally 1-pinnate, leaflets up to 12 cm long; cauline leaves 1-2 pinnate. Inflorescence with peduncle up to 20 cm long. Flower petals 1-veined, whitish. Styles ≥ 1 mm, slender. Flowering time: June-October.
Plant Associates: Carex densa, Clarkia purpurea, Danthonia californica, Deschampsia cespitosa, Eryngium armatum, Juncus sp., Perideridia kelloggii, Ranunculus californicus, Sisyrinchium bellum.
Comments: Endangered in the southern portion of its range; status of occurrences uncertain. Can be relatively common locally, especially in northern counties. Is plant extant in SMT Co.? Threatened by agriculture and urban development. See *University of California Publications in Botany* 55:1-74 (1969) for taxonomic treatment. T258 J160

Apiaceae (carrot) ***Sanicula hoffmannii***
Hoffmann's sanicle
CNPS List: 4 **State/Federal Status:** /C3c
Distribution: MNT, SBA, SCR, SCZ, SLO, SMT, SRO
Habitat: Coastal scrub, coniferous forest, mixed evergreen forest; in coastal areas; often ultramafic or clay.
Description: Stout perennial up to 9 dm tall, with a taproot. Leaf generally compound, palmate, whitish-bluish green; blade up to 13.5 cm long, triangular. Inflorescence with peduncle up to 12 cm long; pedicels 0 or short. Involucel bractlets 3-5 mm long. Flowers greenish yellow, styles about equal to calyx lobes. Fruit covered with hooked prickles. Flowering time: March-May.
Plant Associates: Agrostis pallens, Bromus carinatus, Dryopteris arguta, Festuca californica, Heteromeles arbutifolia, Holodiscus discolor, Melica torreyana, Pseudotsuga menziesii, Pteridium aquilinum pubescens, Quercus wislizenii, Rubus ursinus, Sanicula crassicaulis, Stachys bullata, Toxicodendron diversilobum.
Comments: See: Hoover R.F., *The Vascular Plants of San Luis Obispo County*, 1970. J163

Apiaceae (carrot) ***Sanicula saxatilis***
rock sanicle
CNPS List: 1B **State/Federal Status:** CR/C2
Distribution: CCA, SCL
Habitat: Rocky ridges and talus slopes in chaparral, foothill woodland, valley and foothill grassland.
Description: Perennial herb from 10-25 cm high, with diverging branches arising from a massive, deep-seated tuberous root. Leaves strongly dissected, shining, borne flat on the ground. Flowers small and numerous, pale salmon to straw-colored, borne in small globose clusters. Fruit 2.5-3 mm wide, covered with inflated tubercles, the uppermost bearing curved (but not hooked) prickles. Flowering time: February-May.
Plant Associates: Acanthomintha lanceolata, Allium falcifolium, Amsinckia eastwoodiae, Arabis breweri, Ceanothus sp., Claytonia gypsophiloides, Collinsia sparsiflora, Delphinium nudicaule, Galium aparine, Idahoa scapigera, Lewisia rediviva, Linanthus pygmaeus, Malacothrix sp., Moehringia macrophylla, Nemophila heterophylla, Phacelia breweri, Phacelia distans, Pinus sabiniana, Plectritis congesta, Rigiopappus leptocladus, Thysanocarpus laciniatus.
Comments: Known only from fewer than fifteen occurrences. Threatened by development. See Erythea 1:6 (1893) for original description,and *University of California Publications in Botany* 25:71-72 (1951) for taxonomic treatment. J164

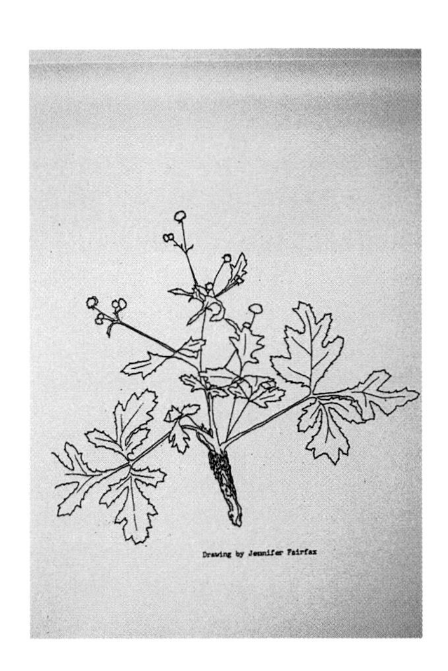
Drawing by Jennifer Fairfax

Asteraceae (sunflower) ***Balsamorhiza macrolepis macrolepis***
big-scale balsamroot
CNPS List: 1B **State/Federal Status:** CEQA
Distribution: ALA, BUT, MPA, NAP, PLA, SCL, TEH
Habitat: Valley and foothill grassland and foothill woodland slopes.
Description: Perennial herb from 2-6 dm tall, finely glandular-hairy. Basal leaves 20-40 cm long; blade lanceolate to elliptic, pinnately divided into linear segments; cauline leaves often 2-3 near base. Inflorescence of 1 head; outer phyllaries 2-4 cm long, generally 6-10 mm wide, ovate-lanceolate, tip wide, flat, obtuse to acute, appressed to spreading, finely strigose, minutely glandular. Ray flowers 2-3 cm long; disk flower corollas 7-10 mm long. Flowering time: March-June.
Plant Associates: Achillea millefolium, Anthriscus caucalis, Avena barbata, Bromus diandrus, Bromus hordeaceus, Castilleja sp., Chlorogalum pomeridianum, Elymus glaucus, Hemizonia congesta luzulifolia, Lolium multiflorum, Lolium perenne, Perideridia sp., Quercus agrifolia, Quercus lobata, Rhamnus californica, Toxicodendron diversilobum.
Comments: See *Annals of the Missouri Botanical Garden* 22:132 (1935) for original description. J213

Asteraceae (sunflower) ***Cirsium andrewsii***
Franciscan thistle
CNPS List: 4 **State/Federal Status:** CEQA?
Distribution: MRN, SFO, SMT, SON
Habitat: On coastal: bluffs, prairie, ravines, scrub and seeps. Sometimes ultramafic.
Description: Branched biennial up to 20 dm. Plant thinly cobwebby, becoming glabrous. Lvs cobwebby above, gray-cobwebby below. Lvs stiff, strongly spiny, spines stout, often 10-15 mm long; most outer phyllaries spiny-margined. Fl bell-shaped, closely subtended by leafy bracts, involucre bracts densely cobwebby; petals to 24 mm long, dark reddish-purple. Anthers exserted beyond the corolla. Flowering time: June-July.
Plant Associates: Angelica sp., Arabis blepharophylla, Circium quercetorum, Conyza sp., Eschscholzia californica, Iris douglasiana, Juncus sp., Mentha pulegium, Mimulus guttatus, Oenanthe sarmentosa, Phacelia californica, Plantago coronopus, Pteridium aquilinum, Rubus ursinus.
Comments: T379 J235

Asteraceae(sunflower) ***Cirsium fontinale campylon***
Mt. Hamilton thistle
CNPS List: 1B **State/Federal Status:** /C2
Distribution: ALA, SCL, STA
Habitat: Ultramafic seeps in valley and foothill grassland
Description: Perennial herb with stout, erect, greenish or purplish tinged, wingless stems up to 2 m tall. Plant leafy throughout, often with a basal rosette. Leaves spiny, upper surface densely tomentose, glandular hairs concealed. Outer phyllaries strongly recurved, 2-3 cm long, spines 2-6 mm long. Flowers white to pink-lavender; heads strongly nodding. Flowering time: April-October.
Plant Associates: Asclepias fascicularis, Bromus hordeaceus, Carex sp., Cirsium vulgare, Cortaderia selloana, Cotula coronopifolia, Cyperus esculentus, Eleocharis macrostachya, Elymus sp., Epilobium sp., Eriogonum nudum, Eschscholzia californica, Juncus balticus, J. phaeocephalus, J. xiphioides, Lactuca sp., Mimulus guttatus, Plantago erecta, Polypogon monspeliensis, Rumex salicifolius, Scirpus sp., Sonchus sp., Verbena lasiostachys scabrida.
Comments: Threatened by urbanization, and by cattle trampling and grazing. See *Phytologia* 73(4):312-317 (1992) for revised nomenclature. Synonym: *C. campylon.* J236

Asteraceae (sunflower) ***Cirsium fontinale fontinale***
fountain thistle
CNPS List: 1B **State/Federal Status:** CE/FE
Distribution: SMT
Habitat: Ultramafic seeps and ravines in valley and foothill grassland.
Description: Perennial herb with stout, erect, reddish, wingless stems up to 1.3 m tall. Leaves spiny, thinly tomentose, clearly glandular hairs on upper surface. Outer phyllaries 1.5-2 cm long, spines 1-2 mm long. Flower heads single or in close clusters and nodding. Flowers whitish to pinkish-lavender, becoming brown with age. Flowering time: June-October.
Plant Associates: Baccharis douglasii, Carex serratodens, Deschampsia cespitosa holciformis, Festuca arundinacea, Juncus effusus pacificus, Juncus xiphioides, Leymus triticoides, Mimulus guttatus, Perideridia kelloggii, Plantago subnuda, Rorippa nasturtium-aquaticum, Stachys ajugoides.
Comments: Known from only four occurrences in the vicinity of Crystal Springs Reservoir. Seriously threatened by urbanization, dumping, road maintenance, and non-native plants. See *Bulletin of the California Academy of Sciences* 2:151-152 (1886) for original description. T378 J236

Asteraceae (sunflower) ***Coreopsis hamiltonii***
Mt. Hamilton coreopsis
CNPS List: 1B **State/Federal Status:** /C2
Distribution: SCL, STA
Habitat: Dry, exposed, rocky slopes in foothill woodland.
Description: Erect, glabrous annual from 10-15 cm tall. Leaves mostly basal, 2-5 cm long, bipinnate. Outer phyllaries linear, shorter than the broad, glabrous inner ones. Flower heads golden, 1-2 cm wide, ray flowers strongly reflexed at anthesis. Flowering time: March-May.
Plant Associates: Acanthomintha lanceolata, Adenostoma fasciculatum, Arabis breweri, Athysanus pusillus, Chaenactis glabriuscula heterocarpha, Erysimum capitatum, Gilia achilleifolia, Isopyrum stipitatum, Linanthus dichotomus, Lotus wrangelianus, Lupinus bicolor, Malacothrix floccifera, Melica bulbosa, Mentzelia lindleyi, Parvisedum pentandrum, Phacelia breweri, Pinus sabiniana, Quercus kelloggii, Rhamnus crocea, Salvia columbariae, Streptanthus callistus, Trifolium albopurpureum, Viola purpurea quercetorum.
Comments: Known from fewer than ten occurrences in the Mt. Hamilton Range. Grazing and competition from non-native plants are threats. See *Botanical Gazette* 41:323-324 (1906) for original description, and *Madrono* 4:214-215 (1938) for revised nomenclature. J241

Asteraceae (sunflower) ***Eriophyllum jepsonii***
Jepson's woolly sunflower
CNPS List: 4 **State/Federal Status:** CEQA?
Distribution: ALA, CCA, KRN, SBT, SCL, STA, VEN
Habitat: Oak woodland and chaparral, sometimes ultramafic.
Description: Perennial subshrub 5-8 dm tall. Leaves 3-6 cm long, ovate, pinnately 5-7-lobed, woolly-tufted, becoming glabrous above. Inflorescence of 1-5 heads, heads < 15 mm diam; peduncles 4-14 cm long; phyllaries acute, strongly overlapping, keeled, free. Ray flowers 6-8; ligules 6-10 mm long. Disk corollas 3-5 mm long, glandular-puberulent or bristly. Flowering time: April-June.
Plant Associates: Aspidotis carlotta-halliae, Calochortus invenustus, Chaenactis glabriuscula, Elymus sp., Fritillaria falcata, Malacothrix californica, Melica sp., Pellaea mucronata, Streptanthus breweri.
Comments: J264

Asteraceae (sunflower) ***Eriophyllum latilobum***
San Mateo woolly sunflower
CNPS List: 1B **State/Federal Status:** CE/FE
Distribution: SMT
Habitat: Ultramafic in oak woodland, exposed grassland roadcuts.
Description: Bushy yellow-flowered perennial up to 5 dm tall. Leaves deeply 3-cleft at the middle and often toothed or lobed, hairy, to 6 cm long. Flowers in loose heads of up to 10 flowers on long peduncles up to 8 cm long. Bright yellow ray flowers to 10 mm long with many yellow disk flowers. Flowering time: April-June.
Plant Associates: Aesculus californica, Artemisia californica, Avena fatua, Calochortus albus, Chlorogalum pomeridianum, Cirsium occidentale, Heteromeles arbutifolia, Mimulus aurantiacus, Nassella lepida, Quercus agrifolia, Toxicodendron diversilobum, Umbellularia californica.
Comments: Known only from one extant occurrence. Threatened by development, erosion and road maintenance. See Carlquist, S. 1956. On the genetic limits of *Eriophyllum* (Composite) and related genera. *Madrono* 13:226-239. J264

Asteraceae (sunflower) ***Grindelia hirsutula maritima***
San Francisco gumplant
CNPS List: 1B **State/Federal Status:** /C2
Distribution: MRN, MNT, SCR, SFO, SLO, SMT
Habitat: Sandy or ultramafic slopes on coastal bluffs, coastal scrub and grassland.
Description: Perennial plant to 5 dm tall, often leaning, with reddish stems. Leaves generally gray-green, to 10 cm long, not deeply cut, not fleshy. Flower bright yellow. Fruit golden to gray-brown, deeply ridged, top generally knobby, pappus awns less than .3 mm wide at base and entire. Flowering time: August-September.
Plant Associates: Avena sp., Baccharis pilularis, Dactylis glomerata, Ericameria ericoides, Erigeron glaucus, Eschscholzia californica, Hordeum brachyantherum, Mesembryanthemum crystallinum, Nassella pulchra, Plantago lanceolata, Scrophularia californica, Sidalcea malvaeflora, Tritelia laxa.
Comments: Need current information on distribution. Threatened by coastal development and non-native plants. See *Pittonia* 2:289 (1892) for original description, and *Novon* 2(3):215-217 (1992) for revised nomenclature. Synonym: *Grindelia maritima*. T346 J274

Asteraceae (sunflower) ***Grindelia stricta angustifolia***
marsh gumplant
CNPS List: 4 **State/Federal Status:** CEQA?
Distribution: ALA, CCA, MNT, MRN, NAP, SCL, SFO, SMT, SOL, SON
Habitat: Tidal areas, coastal saltwater marsh.
Description: Perennial subshrub up to 15 dm tall, erect and glabrous. Stems generally red-brown throughout. Leaves on stem narrower near middle, fleshy. Flowering heads 1 to many generally not subtended by bracts. Flowers 3-5 cm across, bright yellow. Phyllaries slightly recurved. Pappus awns > .3 mm wide at base, minutely serrate. Flowering time: August-October.
Plant Associates: Atriplex sp., Brassica nigra, Frankenia salina, Salicornia virginica.
Comments: Rare in MNT Co. Hybridizes with *Grindelia camporum camporum*. See *Novon* 2(3):215-217 (1992) for revised nomenclature.
J274

Asteraceae (sunflower) ***Helianthella castanea***
Diablo helianthella
CNPS List: 1B **State/Federal Status:** /C2
Distribution: ALA, CCA, MRN*, SFO*, SMT
Habitat: Open grassy sites in chaparral, coastal scrub, riparian woodland, valley and foothill grasslands.
Description: Perennial up to 5 dm, glabrous to coarsely hairy. Few stem leaves. Outer phyllaries enlarged, leaf-like up to 10 cm long, curving up around head. Flowering head generally one, peduncle to 20 cm, stout, rough-hairy. Flowers yellow, anthers yellow. Fruit glabrous. Flowering time: April-May.
Plant Associates: Achillea millefolium, Adenostoma fasciculatum, Baccharis pilularis, Pentagramma triangularis, Perideridia kelloggii, Quercus agrifolia, Symphoricarpos albus laevigatus, Toxicodendron diversilobum, Trillium chloropetalum, Umbellularia californica.
Comments: Threatened by urbanization, grazing and fire suppression.
T358 J277

Asteraceae (sunflower) ***Hemizonia parryi congdonii***
Congdon's tarplant
CNPS List: 1B **State/Federal Status:** /C1
Distribution: ALA*, CCA*, MNT, SCL(*?), SCR*, SLO, SOL*
Habitat: Alkaline soils in valley and foothill grassland. Sumps and disturbed sites where water collects.
Description: Annual herb from 1-7 dm tall, prostrate to erect. Leaves soft-hairy or bristly, not glandular, lower generally absent at flowering, 5-20 cm long, deeply divided; upper linear, entire or few-toothed, spine-tipped. Inflorescence open or dense; heads (sub)sessile; involucre 4.5-8 mm long; ligules 2.5-3 mm, remaining yellow; disk pappus present, anthers yellow. Flowering time: June-November.
Plant Associates: Brassica sp., Briza minor, Centaurea sp., Conyza canadensis, Cirsium vulgare, Heterotheca grandiflora, Madia sp., Picris echioides, Polypogon sp., Rumex maritimus, Senecio californicus, Verbena lasiostachys, Vulpia myuros.
Comments: Nearly extirpated from the San Francisco Bay Area; need information on historical and present distribution. Severely threatened by development. See *Botanical Gazette* 22:169 (1896) for original description, and *Madrono* 3(1):15 (1935) for revised nomenclature. T363 J283

Asteraceae (sunflower) ***Isocoma menziesii diabolica***
Satan's goldenbush
CNPS List: 4 **State/Federal Status:** CEQA?
Distribution: SBT, SCL
Habitat: Open slopes and cliffs in foothill woodland.
Description: Erect shrub 4-6 dm tall, with distinctly whitish stems. Stems, leaves, and phyllaries densely viscid with stipitate glandular hairs. Largest lower leaves 2-4 cm long, 5-10 mm wide, margins shallowly serrate, upper sharply reduced in size, entire. Heads 20-26 flowered. Flowering time: August-October.
Plant Associates: Eriogonum nudum, Chrysothamnus nauseosus mohavensis, Eriodictyon californicum, Isocoma acradenia bracteosa, Juniperus californica, Quercus turbinella.
Comments: Not in *The Jepson Manual.* See *Phytologia* 70(2):69-114 (1991) for original description.

Asteraceae (sunflower) ***Lasthenia conjugens***
Contra Costa goldfields
CNPS List: 1B **State/Federal Status:** /C1
Distribution: ALA*, CCA*, MEN*, NAP, SBA*, SCL*, SOL
Habitat: Vernal pools, moist valley and foothill grassland.
Description: Annual herb, 1-3 dm high, usually branched; leaves opposite, light green, entire to pinnately cut; flowers in heads, terminal, yellow; phyllaries partly fused (1/3 to 1/2 fused); achene less than 1.5 mm long and lacking pappus. Flowering time: March-June.
Plant Associates: Cotula sp., Lasthenia californica, Lasthenia fremontii, Lasthenia platycarpha, Lupinus bicolor, Lepidium sp., Muilla maritima, Triphysaria eriantha.
Comments: Known from only four occurrences after comprehensive 1993 surveys. Many historical occurrences have been extirpated by development and over grazing. J299

Asteraceae (sunflower) ***Lessingia arachnoidea***
Crystal Springs lessingia
CNPS List: 1B **State/Federal Status:** /C2
Distribution: SMT, SON?
Habitat: Open ultramafic barrens, valley and foothill grasslands, coastal scrub and roadsides.
Description: Erect annual to 8 dm high. Herbage tomentose, without glands. Leaves to 11 cm long, lanceolate, entire to toothed, reduced upwards. Involucral bracts tomentose, lacking glands. Flower heads solitary, no ray flowers, disk flowers 5 lobed, pale to deep lavender. Pappus < fruit length. Basal leaves fall off before flowering. Flowering time: July-October.
Plant Associates: Achillea millefolium, Aster chilensis, Avena sp., Centaurea solstitialis, Eriogonum nudum, Madia elegans, Nassella pulchra, Perideridia kelloggii.
Comments: Known only from Crystal Springs Reservoir in SMT Co; occurrences from SON Co. need taxonomic verification. See *Leaflets of Botanical Observation and Criticism* 2:29 (1910) for original description. Synonym: *L. hololeuca var. arachnoidea.* T353 J305

Asteraceae (sunflower) ***Lessingia germanorum***

San Francisco lessingia

CNPS List: 1B **State/Federal Status:** CE/C1

Distribution: SFO, SMT

Habitat: Restricted to sandy soils of remnant dunes and coastal scrub.

Description: Decumbent annual up to 3 dm tall. Stems reddish brown. Basal leaves deciduous, up to 5 cm long, pinnately lobed, without glands. Stem leaves 0.5-3 cm long, oblong, entire to pinnately lobed. No ray flowers, disk flowers 20-40, tubular shaped, deep lemon-yellow with reddish-brown band in throat. Flowering time: August-November.

Plant Associates: Avena sp., Baccharis pilularis, Bromus diandrus, Camissonia micrantha, Chorizanthe cuspidata, Conyza sp., Clarkia purpurea, Croton californicus, Eschscholzia californica, Pteridium aquilinum pubescens, Rumex acetosella, Sonchus oleraceus.

Comments: Known only from four occurrences at the Presidio of San Francisco, and one on San Bruno Mtn., SMT Co. Threatened by urbanization, base-closure activities, trampling and non-native plants. T353 J305

Asteraceae (sunflower) ***Lessingia hololeuca***

woolly-headed lessingia

CNPS List: 3 **State/Federal Status:** CEQA?

Distribution: ALA, MNT, MRN, NAP, SCL, SMT, SOL, SON, YOL

Habitat: Ultramafic, clay soils in coastal scrub, coniferous forests, valley and foothill grasslands.

Description: Erect annual up to 4 dm tall, herbage without glands. Basal leaves persistent, tomentose, up to 5 cm long, linear, entire to lobed, stem leaves 0.2-3 cm long, reduced upwards, lanceolate, entire. Head solitary or in tight clusters, phyllaries woolly, no ray flowers, disk flowers pink to lavender, funnel-shaped. Pappus equal to or greater than length of fruit. Flowering time: June-October.

Plant Associates: Elymus multisetus, Epilobium brachycarpum, Lactuca saligna, Madia elegans, Nassella pulchra.

Comments: Need location, rarity and endangerment information. Probably more widespread. Possibly threatened by grazing. See *Flora Franciscana*, p. 377 (1897) by E.L. Greene for original description, and *UC Publications in Botany* 16:40 (1929) for taxonomic treatment. T353 J305

Asteraceae (sunflower) ***Lessingia micradenia glabrata***
smooth lessingia
CNPS List: 1B **State/Federal Status:** /C2
Distribution: SCL
Habitat: Ultramafic soils in chaparral, valley and foothill grasslands (often roadsides).
Description: Annual herb .5-6 dm tall with spreading branches, thinly tomentose on upper surface of cauline lvs; basal leaves deciduous, to 6 cm long, entire, toothed or lobed; cauline leaves .2-2 cm long, linear to lanceolate, without glands on margins. Flowers white to lavender, solitary or clusters of 3-5, no ray flowers; phyllaries glabrous. Flowering time: August-November.
Plant Associates: Cryptantha flaccida, Dudleya setchellii, Eriogonum nudum, Eriophyllum confertiflorum, Lotus humistratus, Lotus wrangelianus, Nassella pulchra, Salvia columbariae.
Comments: See Aliso 4:105 (1958) for original description, and *Contributions from the Dudley Herbarium* 5:101 (1958) for revised nomenclature. Synonym: *L. ramulosa var. glabrata.* T 354 J306

Asteraceae (sunflower) ***Pentachaeta bellidiflora***
white-rayed pentachaeta
CNPS List: 1B **State/Federal Status:** CE/FE
Distribution: MRN*, SCR*, SMT
Habitat: Ultramafic grasslands.
Description: Annual herb up to 17 cm tall, unbranched or one or two primary branches originating from the base or lower third of plant. Leaves narrowly linear. Ray flowers 5-16, white or purplish-tinged, the ligule 5 mm long; disk flowers numerous, yellow. Newly developed ligules are often pink-purple, turning white after the ligule reflexes. Pappus of five bristles, sometimes lacking. Flowering time: March-May.
Plant Associates: Brodiaea terrestris, Calystegia subacaulis, Castilleja densiflora, Chlorogalum pomeridianum, Delphinium hesperium, Elymus multisetus, Eschscholzia californica, Gilia clivorum, Hesperevax sparsiflora, Lasthenia californica, Layia platyglossa, Linanthus ambiguus, L. androsaceus, Lomatium macrocarpum, Lotus wrangelianus, Microseris douglasii, Nassella pulchra, Plantago erecta, Poa secunda, Sanicula bipinnatifida, Sisyrinchium bellum,Trifolium willdenovii.
Comments: Known only from one extended occurrence bisected by Hwy. 280; historical occurrences lost to development. See *Bulletin of the California Academy of Sciences* 1:86 (1885) for original description, and *University of California Publications in Botany* 65:1-4 (1973) for taxonomic treatment. Synonym *Chaetopappa bellidiflora.* T349 J323

Asteraceae (sunflower) ***Psilocarphus brevissimus multiflorus***
delta woolly-marbles
CNPS List: 4 **State/Federal Status:** CEQA?
Distribution: ALA, NAP, SCL, SJQ, SOL, STA, YOL
Habitat: Vernal pools and flats.
Description: Annual herb generally thinly silky-tomentose. Stems generally branched above, usually erect; central stem dominant. Uppermost leaves lanceolate to ovate from 14-25 mm long. Largest head 9-14 mm, ovoid; receptacle deeply lobed; longest chaff scale greater than 3 times as wide, narrowly cylindric, wing about 1/2 distance from base to tip of scale. Flowering time: May-June.
Plant Associates:
Comments: Does plant occur in CCA, SAC, or other counties? Similar to *P. elatior*. See *Research Studies State College of Washington* 18:80 (1950) for original description. T356 J329

Asteraceae (sunflower) ***Senecio aphanactis***
rayless ragwort
CNPS List: 2 **State/Federal Status:** CEQA
Distribution: CCA, FRE, LAX, MER, ORA, RIV, SBA, SCL, SCZ, SDG, SLO, SOL, VEN, BA
Habitat: Drying alkaline flats in foothill woodland and coastal scrub.
Description: Annual herb, 1-2 dm tall, branched upward, usually glabrous (except heads). Leaves equal, evenly distributed or crowded upward, 2-4 cm long, linear to oblanceolate, subentire. Inflorescence of 4-10+ heads, usually urn-shaped; main phyllaries < 13, 5-6 mm long, tips green. Ray flowers few, ligules yellow, barely longer than phyllaries. Fruit densely gray-hairy. Flowering time: January-April.
Plant Associates:
Comments: Rare in LAX, ORA, and RIV counties. Need quads for RIV Co. Not seen on SCZ Isl. between 1934 and 1991. See *Pittonia* 1:220 (1888) for original description, and *North American Flora* II 10:50-139 (1978) for taxonomic treatment. T374 J337

Azollaceae (mosquito fern) ***Azolla mexicana***
Mexican mosquito fern
CNPS List: 4 **State/Federal Status:** CEQA?
Distribution: BUT, KRN, LAK, MOD, NEV, PLU, SCL, SDG, TUL, AZ, BA, GU, NV, OR, ++
Habitat: Ponds, slow streams, wet ditches, marshes and swamps.
Description: Annual herb, plant green or blue-green to dark red or red-fringed, generally fertile. Stems prostrate, generally 1-2 cm long; internodes up to 1 mm long. Sporangium cases often male and female. Fertile: August.
Plant Associates: Glyceria elata, Hydrocotyle ranunculoides, Scirpus acutus, Typha latifolia.
Comments: Difficult to distinguish from *Azolla filiculoides*. See *American Fern Journal* 34(3):69-84 (1944) for a review of New World Azolla. J90

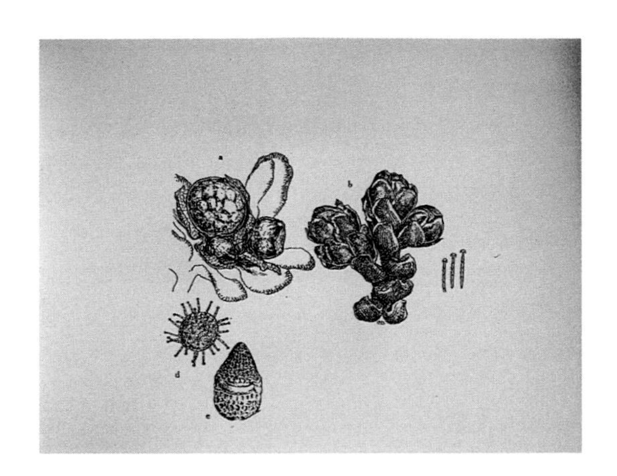

Boraginaceae (borage) ***Plagiobothrys chorisianus chorisianus***
Choris's popcorn-flower
CNPS List: 3 **State/Federal Status:** CEQA?
Distribution: SCR, SFO, SMT
Habitat: Grassy and moist places in coastal scrub and chaparral.
Description: Decumbent to erect annual, branched from upper axils. Lower leaves generally fused at base, up to 7 cm long, loosely sheathing stem. No basal rosette of leaves. Inflorescence with pedicel generally greater than calyx. Corolla 6-10 mm wide. Fruit a nutlet; nutlet scar sessile. Flowering time: April-June.
Plant Associates: Castilleja densiflora, Dudleya sp., Erigeron glaucus, Plantago lanceolata, Plantago maritima, Polypodium californicum, Pteridium aquilinum pubescens, Ranunculus occidentalis.
Comments: Move to List 1B? Taxonomic work needed: differences from var. *hickmanii* may be environmentally induced. Synonym: *Allocarya chorisiana var.chorisiana*. T292 J388

Boraginaceae (borage) ***Plagiobothrys glaber***
hairless popcorn-flower
CNPS List: 1A **State/Federal Status:** /C2
Distribution: ALA*, MER*, MRN*, SBT*, SCL*
Habitat: Wet, alkaline soils in valleys, coastal marshes, meadows, swamps
Description: Annual herb with thick, hollow stem. Lower cauline leaves opposite, upper alternate; no basal rosette, lower 2-11 mm long. Pedicel thick, hollow; calyx 8-10 mm long in fruit, bottom 2-3 mm fused into fleshy cylinder; corolla 3 mm wide. Fruit a nutlet 1.5-2.5 mm long, scar basal, generally on short peg. Flowering time: April-May.
Plant Associates:
Comments: Last seen in 1954. All collections since 1930's located in the Hollister area; plant should be looked for there. A variety or ecotype of *P. stipitatus*? See *Proceedings of the American Academy of Arts and Sciences* 17:227 (1882) for original description. Synonym: *Allocarya glabra*. T291 J388

J

HERBARIUM OF THE
78149

Boraginaceae (borage) ***Plagiobothrys myosotoides***
forget-me-not popcorn-flower
CNPS List: 4 **State/Federal Status:** CEQA?
Distribution: FRE, SCL, TUL, SA
Habitat: Chaparral.
Description: Spreading annual herb with stiff hairs and purple sap. Stem erect, 5 cm-2 dm tall. Basal leaves in loose rosette, 1-2 cm long, stem leaves alternate. Calyx 2.5 mm long, sepal hairs not hooked; corolla 1.5-2 mm wide. Fruit a nutlet 1.5 mm long; scar short, round. Flowering time: April-May.
Plant Associates: Adenostoma fasciculata, Eriophyllum confertiflorum.
Comments: Identification uncertain. Relationship to *P. torreyi* complex needs study. More SA species needed for comparison with CA material. J389 Grows near *Plagiobothrys tenellus* except no purple sap or purple color under leaves and fruit.

Boraginaceae (borage) ***Plagiobothrys uncinatus***
hooked popcorn-flower
CNPS List: 1B **State/Federal Status:** /C2
Distribution: MNT, SBT, SCL, SLO
Habitat: Canyon sides, chaparral (sandy), foothill woodland, valley and foothill grassland.
Description: Spreading annual herb with stiff hairs and purple sap. Stem decumbent to erect, 5cm-2 dm tall. Basal leaves in loose rosette,1-2 cm long; stem leaves alternate. Calyx 2-2.5 mm long, hairs minutely hooked at tip; corolla 1.5-2 mm wide. Fruit a nutlet 1-1.3 mm, widely ovoid, scar lateral near middle, round. Flowering time: May.
Plant Associates:
Comments: Field surveys needed in Gabilan and Santa Lucia ranges to determine status. J390

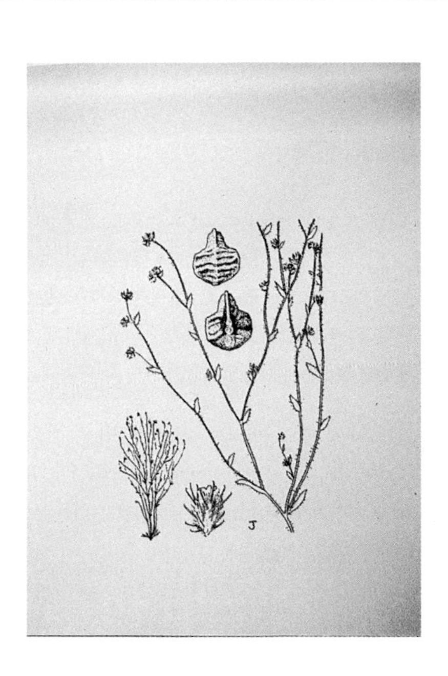

Brassicaceae (mustard) ***Arabis blepharophylla***
coast rock cress
CNPS List: 4 **State/Federal Status:** /C3c
Distribution: CCA, MRN, SCR, SFO, SMT, SON
Habitat: Rocky outcrops, steep banks in coastal: scrub and prairie.
Description: Perennial herb to 2 dm tall with an obvious basal rosette of leaves and some stem leaves. Stem leaves sessile. Flower petals 4, rose-purple, to 12 mm long, fragrant. Fruit a narrow pod 2-4 cm, pedicel erect, 5-20 mm long, less than 3 mm wide, straight, hairy. Flowering time: February-May.
Plant Associates: Acaena pinnatifida californica, Allium dichlamydeum, Crassula connata, Dichelostemma capitatum, Dudleya farinosa, Plantago erecta, Ranunculus californicus, Sanicula arctopoides, Sedum spathulifolium, Viola adunca.
Comments: See *Rhodora* 43(511):348-349 (1941) for taxonomic treatment, and *Contributions from the Gray Herbarium* 204:149-154 (1973) for taxonomic information. T182 J399

Brassicaceae (mustard) ***Erysimum ammophilum***
coast wallflower
CNPS List: 1B **State/Federal Status:** /C2
Distribution: MNT, SCR, SMT, SRO
Habitat: Coastal: dunes, strand, scrub.
Description: Biennial or short-lived perennial up to 6 dm tall. Leaves entire or slightly toothed, basal leaves to 15 cm long, over 2 mm wide; stem leaves wider, especially near flowers. Flowers a rich yellow, petals to 25 mm long, style generally to l mm in fruit. Flowering time: February-June.
Plant Associates: Artemisia pycnocephala, Carpobrotus edulis, Ericameria ericoides, Gnaphalium bicolor, Gnaphalium ramosissimum, Lotus scoparius, Lupinus chamissonis, Phacelia ramosissima.
Comments: Need quads for SRO Isl. SDG Co. occurrences previously included in this species are *E. capitatum capitatum*. Threatened by coastal development. T181 J421

Brassicaceae (mustard) ***Erysimum franciscanum***
San Francisco wallflower
CNPS List: 4 **State/Federal Status:** /C2
Distribution: MRN, SCL, SCR, SFO, SMT, SON
Habitat: Ultramafic out crops and granitic cliffs in grassland and coastal: dunes and scrub.
Description: Biennial to short-lived perennial with simple or few branched stems to 5 dm tall. Basal leaves to 20 cm long, > 3 mm wide, generally coarsely toothed. Petals to 26 mm long, cream-colored to yellow, style gen > l mm in fruit. Fruit strongly ascending, to 13 cm long, 2-4 mm wide, flattened, often tapered to tip and commonly tinged with purple. Flowering time: March-June.
Plant Associates: Artemisia californica, Baccharis pilularis, Castilleja densiflora, Chlorogalum pomeridianum, Clarkia purpurea, Danthonia californica, Delphinium variegatum, Dichelostemma capitatum, Eriophyllum staechadifolium, Fritillaria liliacea, Hesperevax sparsiflora, Hesperolinon congestum, Mimulus aurantiacus, Nassella pulchra, Plantago erecta, Poa secunda.
Comments: Rare and declining in SCR Co. Includes *Erysimum franciscanum crassifolium*. Inland plants approach *Erysimum capitatum*. See *Aliso* 4(1):118-121 (1958) for original description. T181 J422

Brassicaceae (mustard) ***Streptanthus albidus albidus***
Metcalf Canyon jewel-flower
CNPS List: 1B **State/Federal Status:** /FE
Distribution: SCL
Habitat: Ultramafic valley and foothill grassland.
Description: Annual herb 5-12 dm tall with bristly hairs toward the base; strongly glaucous stem and leaves. Flowers in leafless terminal racemes. Petals 4, 8-11 mm long, whitish with light purple veins. Sepals greenish white, the upper 3 fused, the lower free and spreading. Stamens in 3 pairs. Fruit erect, flattened pods, 3-8 cm long. Flowering time: April-June.
Plant Associates: Artemisia californica, Avena barbata, Avena fatua, Calochortus venustus, Cirsium fontinale campylon, Clarkia rubicunda, Delphinium hesperium, Dichelostemma congestum, Dudleya setchellii, Eriogonum sp., Eschscholzia sp., Fritillaria liliacea, Lessingia sp., Malacothrix sp., Mentzelia sp., Phacelia imbricata.
Comments: Known only from fewer than ten extant occurrences. Threatened by residential development and vehicles. See *Pittonia* 1:62 (1887) for original description, and Madrono 14(7):217-248 (1958) for taxonomic treatment. Synonym: *S. glandulosus var. albidus*. T180 J440

Brassicaceae (mustard) ***Streptanthus albidus peramoenus***
most beautiful jewel-flower
CNPS List: 1B **State/Federal Status:** /C1
Distribution: ALA, CCA, SCL
Habitat: Ultramafic valley and foothill grassland.
Description: Annual herb, 2-8 dm tall, simple to much branched, sparsely bristly below. Leaves to 22 cm long; basal narrowly oblanceolate, coarsely dentate, cauline linear-lanceolate, lower usually coarsely dentate, upper usually entire. Calyx bilateral, sepals 5-10 mm long, glabrous, lilac-lavender; petals 8-14 mm long, upper longer than lower, purplish. Fruit spreading to ascending, 4-12 cm. Flowering time: April-June.
Plant Associates: Achillea millefolium, Aphanes occidentalis, Avena barbata, Avena fatua, Calystegia subacaulis, Dudleya setchellii, Eriogonum nudum, Eriophyllum confertiflorum, Eschscholzia californica, Gilia achilleifolia, Gilia tricolor, Koeleria micrantha, Lasthenia californica, Lessingia micradenia glabrata, Lewisia rediviva, Linanthus ambiguus, Linanthus dichotomus, Linanthus liniflorus, Lotus humistratus, Melica torreyana, Nassella pulchra, Plantago erecta, Platystemon californicus, Salvia columbariae.
Comments: Threatened by development and grazing. Historical occurrences need field surveys. Similar plants from SLO need further study. See Kruckeberg, A.R. (1958). The taxonomy of the species complex *Streptanthus glandulosus* Hook. *Madrono* 14:217-248 (1958). Synonym: *S. glandulosus ssp. glandulosus*. T180 J440

Brassicaceae (mustard) ***Streptanthus callistus***
Mt. Hamilton jewel-flower
CNPS List: 1B **State/Federal Status:** /C2
Distribution: SCL
Habitat: Open chaparral and foothill woodland.
Description: Small annual herb 3-6 cm high, bristly toward the base. Leaves sessile and elliptical in shape, not wedge-shaped at base. Plants may be branched. Flowers in terminal, leafless racemes. Sepals green, 5 mm long. Petals 4, purplish with whitish margins, 10 mm long, strongly flared at tip. Stamens in 3 pairs. Pods erect, incurved, flattened, 15-20 mm long. Flowering time: April -May.
Plant Associates: Acanthomintha lanceolata, Bromus madritensis rubens, Cercocarpus betuloides, Chaenactis sp., Coreopsis hamiltonii, Lamarckia aurea, Lasthenia californica, Lewisia rediviva, Linanthus sp., Lotus humistratus, Lotus wrangelianus, Malacothrix floccifera, Mentzelia lindleyi, Mimulus douglasii, Pinus sabiniana, Plantago erecta, Salvia columbariae, Trifolium willdenovii.
Comments: Known only from approximately five occurrences in the Mt. Hamilton Range. See *Madrono* 4:205 (1938) for original description. J442

Brassicaceae (mustard) ***Tropidocarpum capparideum***
caper-fruited tropidocarpum
CNPS List: 1A **State/Federal Status:** /C2*
Distribution: ALA*, CCA*, GLE*, MNT*, SCL*, SJQ*
Habitat: Alkaline soils, low hills, valley and foothill grassland.
Description: An erect slender-branched annual herb; stems pilose, 2-2.5 dm high; leaves 1-5 cm long, pinnately cleft into narrow lobes; flowers in small, loose racemes; petals yellow, obovate, 3-4 mm long, slightly exceeding the sepals; pods oblong 5-20 mm long, turgid, 4-valved and 1-celled. Flowering time: March- April.
Plant Associates: Cressa truxillensis, Distichlis spicata, Frankenia salina, Sporobolus airoides.
Comments: Last seen in 1957. Recent attempts to rediscover this plant have been unsuccessful. J448

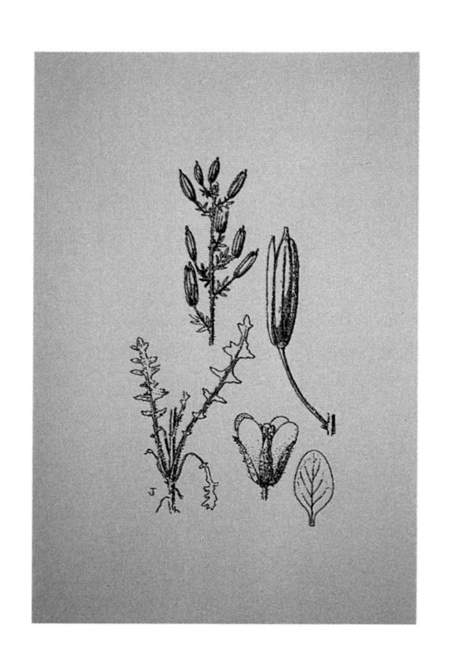

JEPS

Campanulaceae (bellflower) ***Campanula exigua***
chaparral harebell
CNPS List: 4 **State/Federal Status:** CEQA?
Distribution: ALA, CCA, SBT, SCL, STA
Habitat: Talus slopes in clearing of chaparral, often ultramafic.
Description: Annual herb, glabrous or stiffly hairy. Stems erect, 5-20 cm tall. Leaves 5-11 mm long, usually linear, leathery, sessile. Flower pedicel 3-20 mm long; sepals erect; corolla 7-18 mm long, funnel-shaped, lobes spreading, blue; ovary not papillate. Fruit oblong, strongly ribbed; pores near middle. Seed about 0.7 mm long, oblong. Flowering time: May-June.
Plant Associates: Clarkia breweri, Claytonia gypsophiloides, Galium aparine, Gilia tricolor, Pinus sabiniana, Quercus douglasii.
Comments: See *Botanical Gazette* 11:339 (1886) for original description, and *Madrono* 27(4):149-163 (1980) for taxonomic treatment. J 460

Campanulaceae (bellflower) ***Campanula sharsmithiae***
Sharsmith's harebell
CNPS List: 1B **State/Federal Status:** /C2
Distribution: SCL, STA
Habitat: Ultramafic talus slopes in chaparral and foothill woodland.
Description: Annual herb, stiffly hairy. Stem erect, 5-25 cm high. Leaves 5-11 mm long, widely lanceolate, fleshy, serrate, sessile. Flower pedicel 1-3 mm long; sepals erect; corolla 7-16 mm long, funnel- to bell-shaped, deep purple, lobes reflexed; ovary papillate, 2-4.5 mm long. Fruit oblong, papillate, strongly ribbed; pores near middle. Flowering time: May-June.
Plant Associates: Adenostoma fasciculatum, Ceanothus leucodermis, Clarkia breweri, Eriogonum covilleanum, Pinus sabiniana, Quercus durata, Salix breweri, Streptanthus breweri.
Comments: Known only from 3 occurrences. See *Madrono* 27(4):149-163 (1980) for original description. J 460

Campanulaceae (bellflower) ***Legenere limosa***
legenere
CNPS List: 1B **State/Federal Status:** /C2
Distribution: LAK, NAP, PLA, SAC, SMT, SOL, SON*, STA*, TEH
Habitat: Wet areas, vernal pools.
Description: Annual herb growing in moist or wet ground with the base of the plant immersed. Stems lax to 3 dm long. Leaves oblong-lanceolate to 2 cm long, glabrous. Flowers yellowish, single in leaf axils on long, slender pedicels that elongate in fruit to 2-3 times as long as the leaf. Flowering time: May-June.
Plant Associates: Cotula coronopifolia, Eleocharis sp., Pleuropogon californicus.
Comments: Many historical occurrences extirpated. Threatened by grazing and development. See *Pittonia* 2:81 (1890) for original description, *North American Flora* 32(1):13-14 (1943) for revised nomenclature, and *Wasmann Journal of Biology* 33(1-2):91 (1975) for distributional information. Not found in SMT in 1993, may be extirpated from SMT Co. T 331 J464

Caryophyllaceae (pink) ***Silene verecunda verecunda***
San Francisco campion
CNPS List: 1B **State/Federal Status:** /C2
Distribution: SCR, SFO, SMT
Habitat: Sand hills and rocky soils in coastal: strand, prairie, scrub.
Description: Herbaceous perennial from 1-5 dm tall with opposite leaves. Plant densely puberulent. Basal leaves narrowly to broadly oblanceolate, 3-6 cm long, 2-10 mm wide. Inflorescence elongate, flowers with 5 petals, calyx with 5 teeth. Pedicel and calyx short-soft-hairy; flowers pink to rose, petals > 10 mm long. Flowering time: March-August.
Plant Associates: Agoseris apargioides, Delphinium decorum, Dichelostemma capitatum, Dudleya farinosa, Erigeron glauca, Eriogonum latifolium, Eriophyllum staechadifolium, Fritillaria lanceolata, Ranunculus californica, Sanicula arctopoides, Saxifraga californica, Sidalcea malvaeflora, Toxicodendron diversilobum, Viola pedunculata.
Comments: Known only from fewer than twenty occurrences. Threatened by development. See *Proceeding of the American Academy of Arts and Sciences* 10:344 (1875) for original description, and *University of Washington Publications in Biology* 13:41-42 (1947) for taxonomic treatment. T166 J493

Chenopodiaceae (goosefoot) ***Atriplex joaquiniana***
San Joaquin spearscale
CNPS List: 1B **State/Federal Status:** /C2
Distribution: ALA, CCA, COL, GLE, MER, NAP, SAC, SBT, SCL*, SJQ*, SOL, TUL*, YOL
Habitat: Alkaline soils in chenopod scrub, valley and foothill grassland.
Description: Annual herb 1-10 dm high. Stem erect to ascending. Leaves ovate to triangular, finely gray-scaly or green above, 1-7 cm long, irregularly wavy-toothed, upper reduced. Staminate inflorescence spike or panicle-like, terminal, dense. Pistillate inflorescence bracts in fruit 3-4 mm long, free, fused below middle, entire. Flowering time: April-September.
Plant Associates: Allenrolfea occidentalis, Atriplex coronata, Atriplex fruticulosa, Avena barbata, Bromus diandrus, Bromus hordeaceus, Cordylanthus palmatus, Distichlis spicata, Hesperevax caulescens, Frankenia salina, Grindelia sp., Hemizonia pungens, Holocarpha virgata, Hordeum depressum, Hordeum marinum gussoneanum, Lasthenia platycarpha, Lepidium dictyotum acutidens, Lepidium latipes, Lepidium nitidum, Lolium multiflorum, Salicornia subterminalis, Salix laevigata, Spergularia macrotheca, Suaeda sp.
Comments: Need historical quads for TUL Co. Threatened by grazing, agriculture, and development. Synonym: *Atriplex patula ssp. spicata.* T150 J504

Chenopodiaceae (goosefoot) ***Suaeda californica***
California seablite
CNPS List: 1B **State/Federal Status:** /PE
Distribution: ALA*, SCL*, SLO
Habitat: Margins of coastal salt marshes, upper intertidal marsh zone.
Description: Shrub 3-8 dm tall, glabrous or sparsely hairy. Stems decumbent, dull gray-brown; branches spreading, pale green or reddish. Leaves overlapping, petioles 1 mm long forming prominent knobs on lower stems; blades 5-35 mm long, lanceolate, subcylindric to flat. Inflorescences scattered throughout the plant; flowers 1-5 per cluster; bracts equal to the leaves, overlapping at branch tips. Flowers bisexual or pistillate, 2-3 mm long. Flowering time: July-October.
Plant Associates: Atriplex patula, Cakile maritima, Carpobrotus edulis, Cordylanthus maritimus maritimus, Cuscuta salina, Distichlis spicata, Frankenia salina, Jaumea carnosa, Juncus acutus leopoldii, Salicornia virginica.
Comments: Formerly known from the San Francisco Bay Area, where extirpated by development; now extant only in Morro Bay. Often confused with *Suaeda esteroa* and *Suaeda taxifolia* in southern California, but does not occur there. T151 J515

Crassulaceae (stonecrop) ***Dudleya setchellii***
Santa Clara Valley dudleya
CNPS List: 1B **State/Federal Status:** /FE
Distribution: SCL
Habitat: Ultramafic, rocky outcrops in valley and foothill grassland.
Description: Perennial herb, caudex of stem 10-20 mm wide; branches few. Leaves lanceolate to oblong-lanceolate to oblong-triangular, 3-8 cm long, 7-15 mm wide. Terminal inflorescence branches 3-12 cm, primary branches of inflorescence 2-3, generally simple, ascending. Sepals 2-5 mm long, deltate; petals 8-13 mm long, fused for 1-2.5 mm, elliptic, pale yellow, keel not red or purple, tip acute. Flowering Time: May-June.
Plant Associates: Achillea millefolium, Artemisia californica, Avena barbata, Bromus hordeaceus, Calochortus venustus, Claytonia spathulata, Cryptantha flaccida, Dichelostemma congestum, Eschscholzia californica, Gilia tricolor, Hesperevax sparsiflora, Hordeum murinum leporinum, Koeleria micrantha, Lasthenia californica, Lessingia micradenia glabrata, Lewisia rediviva, Linanthus ambiguus, Lotus wrangelianus, Nassella pulchra, Phacelia distans, Plantago erecta, Poa tenerrima, Streptanthus albidus albidus, S. a. peramoenus.
Comments: Known from fewer than 15 occurrences in the Santa Clara Valley. Threatened by urbanization, vehicles and grazing. Synonym: *Dudleya cymosa ssp.setchellii.* T189 J530

Cupressaceae (cypress) ***Cupressus abramsiana***
Santa Cruz cypress
CNPS List: 1B **State/Federal Status:** CE/FE
Distribution: SCR, SMT
Habitat: Sandstone or granitic cypress forest. Stands are separated by coniferous and mixed evergreen forest.
Description: Coniferous tree with a compact, symmetrical, pyramidal crown with branches nearly to ground level. Bark persistent, fibrous, thin, gray, broken in thick vertical strips or plates. Leaves light bright green with a closed dorsal pit. Seed cone spheric up to 30 mm long, brown. Flowering time: March.
Plant Associates: Adenostoma fasciculatum, Arctostaphylos sp., Ceanothus cuneatus, Dendromecon rigida, Ericameria ericoides, Erysimum teretifolium, Pinus attenuata, Pinus ponderosa, Quercus chrysolepis, Quercus wislizenii frutescens.
Comments: Known only from fewer than ten occurrences. Threatened by development, agriculture, and alteration of fire regimes, and possibly by introgression from planted *C. macrocarpa*. Largest known specimen was cut down in 1983. See *Aliso* 1:215-222 (1948) for original description, and *Madrono* 2(4):189-194 (1952) for distributional information. T64 J112

Equisetaceae (horsetail) ***Equisetum palustre***
marsh horsetail
CNPS List: 3 **State/Federal Status:** CEQA?
Distribution: LAK, SFO, SMT, OR, ++
Habitat: Freshwater marshes and swamps.
Description: Perennial from rhizome. Above ground stem annual, of one kind, up to 8 dm, green and firm; basal internode of branch less than the length of the subtending nodal sheath; sheath 4-9 mm long, longer than wide; sheath teeth 5-10, 2-5 mm long, separate; branch with 4-6 rounded ridges, hollow.
Comments: Move to List 2? Location, rarity, and endangerment information needed; need quads. Scarcity poorly understood. Hybrid with *Equisetum telmateia braunii* known from San Mateo County. J95

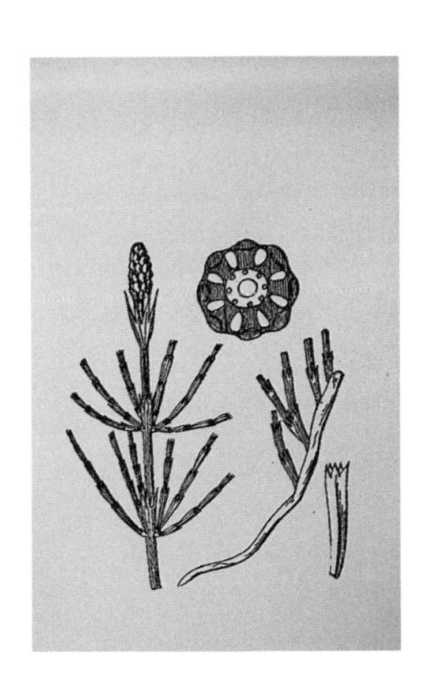

Ericaceae (heath) ***Arctostaphylos andersonii***
Santa Cruz manzanita
CNPS List: 1B **State/Federal Status:** /C2
Distribution: SCL, SCR, SMT
Habitat: Open sites and edges of chaparral, coniferous and evergreen forests.
Description: Erect, evergreen shrub to 5 m tall without a basal burl. Stems with long, white, gland-tipped bristles. Leaves 4-7 cm long, overlapping, clasping, base deeply lobed, leaf surfaces not alike, upper glabrous, convex, lower midrib bristly. Flowers in clusters, white to pink, urn-shaped, ovary finely glandular-bristly. Fruit 6-8 mm wide, finely glandular-bristly, sticky. Flowering time: November-April.
Plant Associates: Arbutus menziesii, Arctostaphylos nummularia , Arctostaphylos silvicola, Arctostaphylos tomentosa, Ceanothus papillosus, Pinus attenuata, Quercus sp.
Comments: Known only from the Santa Cruz Mtns. Has been confused with other species merged with it as varieties. Threatened by development. See *Proceedings of the American Academy of Arts and Sciences* 11:83 (1876) for original description, and *North American Flora* 29:98 (1914) for additional information. T267 J550

Ericaceae (heath) ***Arctostaphylos imbricata***
San Bruno Mtn. manzanita
CNPS List: 1B **State/Federal Status:** CE/C1
Distribution: SMT
Habitat: Chaparral, rocky slopes.
Description: Mat to mound-like evergreen shrub less than 1 m tall, without a basal burl. Stems with long, gland-tipped hairs. Leaves to 4 cm long, strongly overlapping, clasping, heart-shaped, bright green, sparsely hairy, not sticky, leaf surfaces alike, more or less flat. Flowers in clusters, white, urn-shaped, ovary finely glandular-bristly. Fruit 7 mm wide, glandular-hairy, sticky. Flowering time: February-May.
Plant Associates: Ceanothus thyrsiflorus, Rhamnus californica, Vaccinium ovatum.
Comments: Known only from fewer than five occurrences near San Bruno Mtn. Threatened by urbanization and changes in fire regime. See *Proceedings of the American Academy of Arts and Sciences* 20:149-150 (1931) for original description. Synonym: *Arctostaphylos andersonii var. imbricata*. T267 J552

Ericaceae (heath) ***Arctostaphylos montaraensis***
Montara manzanita
CNPS List: 1B **State/Federal Status:** /C2
Distribution: SMT
Habitat: Slopes and ridges in maritime chaparral and coastal scrub on decomposed granitic soil.
Description: Evergreen tree-like shrub to 5 m tall, without a burl. Twigs with long, gland-tipped hairs. Leaves overlapping, light green, to 4.5 cm long, clasping, base lobed, sparsely hairy, not sticky. Pedicel in fruit 3-6 mm. Flowers in dense clusters, pink to white, urn-shaped to 9 mm long, ovary finely glandular-bristly. Fruit 7 mm wide, glandular-hairy. Flowering time: January-March.
Plant Associates: Baccharis pilularis, Ceanothus thyrsiflorus, Garrya elliptica, Gaultheria shallon, Mimulus aurantiacus, Vaccinium ovatum.
Comments: Known only from approximately ten occurrences from San Bruno to Montara Mountain. Threatened by development and vehicles. Synonym: *Arctostaphylos imbricata ssp. montaraensis.* J554

Ericaceae (heath) ***Arctostaphylos regismontana***
Kings Mtn. manzanita
CNPS List: 4 **State/Federal Status:** CEQA?
Distribution: SCR, SMT
Habitat: Granite or sandstone outcrops in chaparral, coniferous and evergreen forests.
Description: Evergreen shrub to 4 m tall without a burl. Leaves strongly overlapping, up to 6 cm long, base clasping, deeply lobed, glandular-hairy, sticky, gray-green, becoming hairless. Pedicel in fruit 6-15 mm. Flowers in clusters, urn-shaped to 9 mm long, ovary finely glandular-bristly. Fruit to 8 mm wide, glandular-hairy, sticky. Flowering time: March-April.
Plant Associates: Adenostoma fasciculatum, Arctostaphylos tomentosa crinita, Baccharis pilularis, Ceanothus thyrsiflorus, Eriodictyon californicum, Lepechinia calycina, Rhamnus californica.
Comments: See *Leaflets of Western Botany* 1:77 (1933) for original description. J556

Fabaceae (pea) ***Astragalus tener tener***
alkali milk-vetch
CNPS List: 1B **State/Federal Status:** CEQA
Distribution: ALA*, CCA*, MER, MNT*, NAP*, SBT*, SCL*, SFO*, SJQ*, SOL, SON*, STA*, YOL
Habitat: Alkaline flats, vernal pools, playas, valley and foothill grassland (adobe clay).
Description: Delicate annual herb, sparsely strigose to glabrous. Stems erect or ascending, 4-30 cm high. Leaves 2-9 cm long, well separated; leaflets 3-16 mm long, lanceolate to obovate, glabrous on upper surface, tip notched or pointed. Inflorescence dense, of 3-12 flowers, spreading. Flowers 7-9 mm long, pink-purple. Fruit reflexed, several times as long as broad, 10-25 mm long, more or less incurved. Flowering Time: March-June.
Plant Associates: Deschampsia danthonioides, Distichlis spicata, Lasthenia glaberrima, Lasthenia platycarpha.
Comments: Last Bay Area collection in 1959. Threatened by habitat destruction, especially agricultural conversion, and protected only at Jepson Prairie Preserve (TNC), SOL Co. See *Proceeding of the American Academy of Arts and Sciences* 6:206 (1864) for original description, and *Systematic Botany* 17(3):367-379 (1992) for distributional information. T217 J604

Fabaceae (pea) ***Lathyrus jepsonii jepsonii***
Delta tule pea
CNPS List: 1B **State/Federal Status:** /C2
Distribution: ALA, CCA, FRE, MRN, NAP, SAC, SBT, SCL, SJQ, SOL
Habitat: Freshwater and brackish marshes.
Description: Perennial herb from underground rootstocks. Semi-erect or prostrate vine-like stems sometimes growing in tangled masses (10-25 dm tall) with broadly winged margins along the internode. Plant parts glabrous. Leaves compound with 10-14 lance like to semi-elliptical leaflets, a terminal tendril, and small stipules. Racemes with 10-20 rose-purple flowers, each 2 cm long. Pea pods 5-9 cm long. Flowering Time: May-June.
Plant Associates: Alnus rhombifolia, Artemisia douglasiana, Aster chilensis lentus, Cephalanthus occidentalis californicus, Cordylanthus mollis mollis, Cortaderia sp., Dipsacus fullonum, Equisetum sp., Hibiscus lasiocarpus, Lepidium latifolium, Phragmites sp., Platanus racemosa, Quercus lobata, Rosa californica, Rubus discolor, Rubus ursinus, Rumex sp., Salix lasiolepis, Scirpus acutus, Scirpus californicus, Triglochin sp.
Comments: Threatened by agriculture and water diversions. See Pittonia 2:158 (1890) for original description. J613

Fabaceae (pea) ***Lupinus eximius***

San Mateo tree lupine

CNPS List: 3 **State/Federal Status:** /C2

Distribution: SMT

Habitat: Coastal: bluffs, scrub.

Description: Silver-hairy shrub to 20 dm tall. Stem leaves with stipules to 12 mm long, petiole less than 6 cm long, leaflets 5-12, to 60 mm long. Inflorescence to 10 cm tall, peduncle 4-10 cm, pedicel 4-10 mm. Flowers whorled, bracts 8-10 mm long, deciduous. Calyx upper lip 5-9 mm long, 2-toothed, lower lip 5-7 mm, entire. Banner mainly yellow, back glabrous, wings blue. Flowering time: April-July.

Plant Associates: Artemisia californica, Baccharis pilularis, Ceanothus thyrsiflorus, Mimulus aurantiacus, Rhamnus californica, Scrophularia californica, Toxicodendron diversilobum.

Comments: Move to List 1B? SON Co. plants need taxonomic confirmation. Identification is very difficult; study needed. See *L. arboreus* in *The Jepson Manual.* USFWS uses the name *L. arboreus var. eximius*. See *Erythea* 3:116 (1895) for original description. T205 J627

Fabaceae (pea) ***Trifolium amoenum***

showy Indian clover

CNPS List: 1B **State/Federal Status:** /C2*

Distribution: ALA*, MEN*, MRN*, NAP*, SCL*, SOL*, SON

Habitat: Moist, heavy soils and disturbed areas in valley and foothill grassland (sometimes ultramafic).

Description: Coarse annual herb 1-6 dm tall, covered by long hairs. Leaves palmately compound with 3 broadly wedge-shaped leaflets with rounded tips. Large, single, globe-shaped flower heads of showy light rose-purple pea flowers with white tips on long leafless stalks above the leaves. Flowers 12-15 mm long, sessile. Calyx with 10 nerves (veins). Flowering time: April-June.

Comments: Rediscovered in 1993 near Bodega Head; only 1 plant found. Habitat lost to urbanization and agriculture. See *Flora Franciscana*, P. 27 (1891) by E. Greene for original description. J649

Hydrophyllaceae (waterleaf) ***Phacelia phacelioides***
Mt. Diablo phacelia
CNPS List: 1B **State/Federal Status:** /C2
Distribution: ALA?, CCA, SBT, SCL, STA
Habitat: Open, rocky slopes in chaparral and foothill woodland.
Description: Annual herb 5-20 cm tall. Stem sparsely stiff-hairy. Leaves 20-80 (100) mm long; elliptic to lanceolate, entire. Flower pedicel 1-3 mm long; calyx lobes 4-6 mm long, densely stiff-hairy; corolla 4-6 mm long, narrowly bell-shaped, white to lavender, lobes violet-streaked, deciduous. Flowering time: April-May.
Plant Associates: Adenostoma fasciculatum, Allium falcifolium, Antirrhinum multiflorum, Apiastrum angustifolium, Bromus madritensis rubens, Calandrinia ciliata, Ceanothus cuneatus, Chorizanthe membranacea, Dicentra chrysantha, Emmenanthe penduliflora, Eriodictyon californicum, Eriogonum covilleanum, Eriogonum nudum, Guillenia lasiophylla, Juniperus californica, Lewisia rediviva, Nemophila heterophylla, Phacelia breweri, Phacelia distans, Pinus sabiniana, Salvia columbariae, Streptanthus hispidus, Trifolium willdenovii.
Comments: Does plant occur in ALA Co.? Possibly threatened by foot traffic and trail construction. J704

Lamiaceae (mint) ***Acanthomintha duttonii***
San Mateo thorn-mint
CNPS List: 1B **State/Federal Status:** CE/FE
Distribution: SMT
Habitat: Ultramafic grassland.
Description: Low, generally unbranched, annual herb with 4-angled stems to 20 cm tall. Leaves bright green, opposite, ovoid shaped, to 12 mm long. Inflorescence bracts 5-11 mm, spine-tipped. Flowers in tight clusters, tubular, to 16 mm long, multicolored white, rose and lavender; style glabrous, anthers pink-red. Flowering time: April-July.
Plant Associates: Agoseris heterophylla, Calochortus albus, Castilleja exserta, Castilleja rubicundula lithospermoides, Delphinium variegatum, Holocarpha virgata, Lolium multiflorum, Lotus micranthus, Lotus wrangelianus, Nassella pulchra, Sidalcea malvaeflora, Trifolium fucatum.
Comments: Known from only two extant natural occurrences; most historical occurrences have been extirpated. Seriously threatened by development. See *Illustrated Flora of the Pacific States* 3:635 (1951) by L. Abrams for original description, and *Madrono* 38(4):278-286 (1991) for revised nomenclature. Synonym: *Acanthomintha obovata ssp. duttonii.* T296 J713

Lamiaceae (mint) ***Acanthomintha lanceolata***
Santa Clara thorn-mint
CNPS List: 4 **State/Federal Status:** CEQA?
Distribution: ALA, FRE, MER, MNT, SBT, SCL, STA
Habitat: Rocky clearings in chaparral, sometimes ultramafic.
Description: Soft hairy annual herb up to 30 cm tall. Stem with hairs short below, conspicuously glandular above. Leaves glandular; blades 10-20 mm long, ovate, upper spiny; margins entire, serrate or spiny. Inflorescence bracts 9-12 mm long, oblong, spines to 12 mm long. Sepals 12 mm long, petals to 2.5 cm long, white to pink-tipped, glandular. Style hairy. Flowering time: March-June.
Plant Associates: Clarkia breweri, Lasthenia californica, Pinus sabiniana, Quercus douglasii, Sanicula saxatilis.
Comments: Threatened by grazing, and by proposed reservoir in MER Co. J713

Lamiaceae (mint) ***Monardella antonina antonina***
San Antonio Hills monardella
CNPS List: 3 **State/Federal Status:** /C3c
Distribution: ALA?, CCA?, MNT, SBT?, SCL?
Habitat: Open, rocky slopes, oak woodland, chaparral.
Description: Open perennial to 6 dm. Leaf lanceolate to narrowly ovate, glabrous or stiff-hairy, lower surface not woolly, conspicuously gland-dotted. Flower head less than 15 mm wide, middle bracts erect, in cup-like involucre. Flowers lavender; sepals hairy. Flowering time: June-August.
Plant Associates:
Comments: Move to List 4? Easily confused with *M. villosa. ssp. villosa*, which may be the taxon occurring in ALA, CCA, SBT, and SCL counties; needs clarification. J719

Lamiaceae (mint) ***Monardella undulata***
curly-leaved monardella
CNPS List: 4 **State/Federal Status:** CEQA?
Distribution: MNT, MRN, SBA, SCR, SFO, SLO, SMT, SON
Habitat: Inland marine sand deposits in coastal: scrub, dunes and chaparral (ponderosa pine sandhills).
Description: Simple or branched, erect annual up to 5 dm tall. Leaves in axillary clusters, to 4 cm long, linear with undulate-wavy margin. Inflorescence a head to 3 cm wide; bracts elliptic to ovate, can be purple-tinged or not. Flower petals to 2 cm long, purple. Flowering time: June-July.
Plant Associates: Adenostoma fasciculatum, Arctostaphylos tomentosa crustacea, Arctostaphylos silvicola, Pinus attenuata, Pinus ponderosa, Salvia mellifera.
Comments: Threatened by coastal development, sand mining and non-native plants. T297 J722

Lamiaceae (mint) ***Monardella villosa globosa***
round-headed coyote-mint
CNPS List: 1B **State/Federal Status:** CEQA
Distribution: ALA, CCA, HUM, LAK ,MRN, NAP, SMT, SON
Habitat: Openings in oak woodland and chaparral.
Description: Erect, open perennial greater than 5 dm tall. Hairs on stem unbranched. Leaf blade 22-50 mm long, base tapered to obtuse. Flower head 2-4 cm wide, outer bracts 2-3 cm, outer reflexed. Flower purple. Flowering time: June-July.
Comments: Need quads for LAK, MRN and NAP Cos. See *Pittonia* 5:82 (1902) for original description, and *Phytologia* 72(1):9-16 (1992) for revised nomenclature. J722

Liliaceae (lily) ***Allium sharsmithae***
Sharsmith's onion
CNPS List: 1B **State/Federal Status:** CEQA
Distribution: ALA, SCL, STA
Habitat: Openings in ultramafic chaparral on talus slopes.
Description: Perennial herb 4-17 cm tall, from a bulb. Leaves about twice as long as the stem, cylindric. Flowers 5-50 per inflorescence, pedicels 6-19 mm. Flowers 10-18 mm long, deep red-purple, perianth parts erect, lanceolate, entire, tips recurved; ovary crests 6, prominent, entire, generally papillate. Flowering time: March-May.
Plant Associates: Fritillaria falcata, Calochortus albus, Calochortus invenustus, Campanula exigua, Clarkia breweri, Eriogonum umbellatum, Eriogonum vimineum, Gilia capitata, Melica sp., Claytonia gypsophiloides, Phacelia sp., Streptanthus breweri.
Comments: Known only from the Mt. Hamilton Range. Synonym: *A. fimbriatum var. sharsmithae*. J1178

Liliaceae (lily) ***Calochortus umbellatus***
Oakland star-tulip
CNPS List: 4 **State/Federal Status:** CEQA?
Distribution: ALA, CCA, MRN, SCL, SCR*, SMT
Habitat: Often ultramafic in: chaparral, grassland, woodland and coniferous forests.
Description: Perennial herb from a bulb. Stem up to 25 cm tall, branched. Leaves mostly basal, one stem-leaf. Inflorescence umbel-like with 3 to 12, bell-shaped flowers; sepals and petals white to pale pink, often with purple spots near base. Petals not conspicuously ciliate, except at the base on the upper surface. Flowering time: March-May.
Plant Associates: Arctostaphylos glauca, Aster radulinus, Cirsium quercetorum, Nassella pulchra, Quercus agrifolia, Quercus berberidifolia.
Comments: Protected at Ring Mtn. Preserve (TNC) on Tiburon Peninsula, MRN Co. T119 J1188

Liliaceae (lily) ***Fritillaria agrestis***
stinkbells
CNPS List: 4 **State/Federal Status:** /C3c
Distribution: ALA,CCA,FRE,KRN,MEN,MNT,MPA,PLA,SAC,SBA, SBT,SLO,SMT,STA,TUO
Habitat: Clay depressions or ultramafic soil in chaparral, valley and foothill woodland.
Description: Perennial from a bulb, 3-6 dm tall. Leaves 5-12, alternate, crowded near lower center of stem, up to 15 cm long. Flowers nodding, odor unpleasant, up to 3.5 cm long, greenish-white or yellow outside, purplish-brown inside, nectary prominent, narrowly linear, green. Flowering time: March-April.
Plant Associates: Avena barbata, Brassica sp, Brodiaea terrestris, Bromus diandrus, Bromus hordeaceus, Centaurea solstitialis, Dichelostemma capitatum, Distichlis spicata, Erodium botrys, Hordeum sp., Isocoma menziesii, Juniperus californica, Lepidium nitidum, Quercus turbinella, Vulpia sp.
Comments: Known locally from along the coast at Ano Nuevo Point and between Santa Cruz and Soquel. Threatened by grazing and development. Synonym: *F. biflora var. agrestis.* T119 J1195

Liliaceae (lily) ***Fritillaria biflora ineziana***
Hillsborough chocolate lily
CNPS List: 1B **State/Federal Status:** CEQA
Distribution: SMT
Habitat: Ultramafic grassland.
Description: Perennial from a bulb, 1 to 4.5 dm tall. Leaves 3-7, alternate, often crowded just above ground level, from 5-19 cm long, 3-6 mm wide. Flower nodding, odor unpleasant, from 1.5 to 2 cm long, dark brown, greenish purple, or yellowish green; nectary prominent, narrowly linear, purplish to greenish. Flowering time: April-May.
Plant Associates: Brodiaea terrestris, Delphinium variegatum, Deschampsia caespitosa holciformis, Dodecatheon clevelandii patulum, Erysimum franciscanum, Muilla maritima, Ranunculus californicus, Sisyrinchium bellum.
Comments: Endemic to the Hillsborough area. See *Flora of California* 1(6):306-307 (1922) by W.L. Jepson for original description. J1195

Liliaceae (lily) ***Fritillaria falcata***
talus fritillary
CNPS List: 1B **State/Federal Status:** /C2
Distribution: ALA, MNT, SBT, SCL, STA
Habitat: Ultramafic talus in chaparral and foothill woodland.
Description: Perennial from a bulb, 1-2 dm tall. Leaves 2-5, glaucous, sickle-shaped, mostly basal, 4-9 cm long, more or less succulent. Perianth clearly mottled, flowers 1-5, erect on ascending pedicels, bowl-shaped, yellowish-green mottled with rust-red inside,15-22 mm long, 5-7 mm wide. Anthers red, style 3-cleft to base, branches strongly recurved. Flowering time: March-May.
Plant Associates: Allium diabloense, Allium falcifolium, Allium sharsmithae, Aspidotis carlotta-halliae, Calochortus invenustus, Cardamine californica, Chaenactis glabriuscula, Clarkia breweri, Collinsia heterophylla, Delphinium nudicaule, Eriogonum umbellatum, Eriophyllum jepsonii, Erysimum capitatum, Fritillaria affinis, Gilia capitata, Juniperus californica, Malacothrix californica, Pellaea mucronata, Phacelia imbricata, Pinus sabiniana, Streptanthus glandulosus secundus.
Comments: Threatened by vehicles (ORVs), mining, and cattle grazing. See *Flora of California* 1(6):309 (1922) by W.L. Jepson for original description, and *Madrono* 7:133-159 (1944) for revised nomenclature. J1195

Liliaceae (lily) ***Fritillaria liliacea***
fragrant fritillary
CNPS List: 1B **State/Federal Status:** /C2
Distribution: ALA, CCA, MNT, MRN, SBT, SCL, SFO, SMT, SOL, SON
Habitat: Moist areas, often ultramafic, open hills, in valley and foothill grasslands.
Description: Herbaceous perennial from an underground bulb,1 to 3.5 dm tall. Lower leaves mostly basal, opposite and somewhat succulent. Upper leaves alternate and smaller. Flower with 6 perianth segments (3 sepals and 3 petals that look alike). Flower color, white with green or yellow throat, fading pinkish at maturity. Flowering time: February-April.
Plant Associates: Camissonia ovata, Chlorogalum pomeridianum, Clarkia purpurea, Dichelostemma capitatum, Hesperevax sparsiflora, Muilla maritima, Nassella pulchra, Plantago erecta, Ranunculus californicus, Sidalcea malvaeflora, Sisyrinchium bellum.
Comments: Threatened by grazing, loss of habitat to agriculture and urban development. J1196

Liliaceae (lily) ***Lilium maritimum***
coast lily
CNPS List: 1B **State/Federal Status:** /C1
Distribution: MEN, MRN*, SFO*?, SMT*, SON
Habitat: Bogs, gaps in closed-cone pine forest, coastal: prairie and scrub.
Description: Perennial from a bulb, to 3 m tall. Leaves basal, scattered or in 1-4 whorls, 3-18 cm long. Flowers 1-13, nodding, bell-shaped, perianth strongly recurved or rolled, red to red-orange, darker spots at mid-base, surrounded by light orange to yellow-green, perianth segments 3-5 cm long. Flowering time: June-July.
Plant Associates: Gaultheria shallon, Iris douglasiana, Lithocarpus densiflorus, Myrica californica, Pteridium aquilinum pubescens, Sisyrinchium bellum, Vaccinium ovatum.
Comments: Did this plant occur in SFO Co.? Populations along Hwy. 1 are routinely disturbed by road maintenance; also threatened by urbanization and horticultural collecting. Hybridizes with *L. pardalinum ssp. pardalinum.* See *Proceedings of the American Academy of Arts and Sciences* 6:140 (1875) for original description. J1199

Limnanthaceae (meadowfoam) ***Limnanthes douglasii sulphurea***
Point Reyes meadowfoam
CNPS List: 1B **State/Federal Status:** CE/C2
Distribution: MRN, SMT
Habitat: Meadows, freshwater marshes, vernal pools and coastal prairie.
Description: Annual to 5 dm tall. Leaves 2-25 cm long. Leaflets ovate, irregularly toothed or lobed. Sepals 10-15 mm long, glabrous. Flowers shallowly cup-shaped to bell-shaped, petals 12-18 mm long, notched at apex, completely and uniformly yellow with scattered long hairs on veins and 2 rows of short hairs at base. Flowering time: March-May.
Plant Associates: Alnus sp., Brassica sp., Bromus sp., Carex sp., Equisetum sp., Holcus lanatus, Juncus sp., Lupinus sp., Potentilla anserina pacifica, Ranunculus californicus, Rubus sp, Scirpus sp., Scrophularia californica.
Comments: Known from approximately ten occurrences. Threatened by grazing, trampling, and non-native plants. See *University of California Publications in Botany* 25:477 (1952) for original description. J738

Linaceae (flax) ***Hesperolinon congestum***
Marin western flax
CNPS List: 1B **State/Federal Status:** CT/FT
Distribution: MRN, SFO, SMT
Habitat: Ultramafic soils in grassland and chaparral.
Description: Annual to 15 cm tall. Stems slender and branched. Leaves long, linear and alternate. Flowers in congested clusters, pedicels 1-8 mm long. Flowers with five rose to whitish petals 3 to 8 mm long; 5 anthers lavender to deep purple at the time of pollen release. Sepals hairy and glandular outside. Fruit an ovoid capsule that produces narrow, 1.5 mm long seeds. Flowering time: May-July.
Plant Associates: Castilleja densiflora, Chlorogalum pomeridianum, Clarkia rubicunda, Delphinium variegatum, Elymus multisetus multisetus, Fritillaria liliacea, Lasthenia californica, Layia platyglossa, Linanthus ambiguus, Lomatium macrocarpum, Nassella pulchra, Plantago erecta, Sisyrinchium bellum.
Comments: Known only from fewer than twenty occurrences. Protected in part at Ring Mtn. Preserve (TNC), MRN Co. Threatened by development and foot traffic. See *Proceedings of the American Academy of Arts and Sciences* 6:521 (1865) for original description, and *University of California Publications in Botany* 32:235-314 (1961) for taxonomic treatment. T226 J739

Malvaceae (mallow) ***Malacothamnus arcuatus***
arcuate bush mallow
CNPS List: 4 **State/Federal Status:** CEQA?
Distribution: SCL, SCR, SMT
Habitat: Ultramafic chaparral.
Description: Upright gray-green shrub to 5 m tall. Leaf blade 2-6 cm long, thin; lower leaf surface covered with dense shaggy-tomentose stellate hairs, upper surface more sparsely hairy. Inflorescence spike-like, many-flowered with sessile flowers. Calyx to 10 mm long, densely stellate hairy. Petals 15-20 mm long, rose-colored. Flowering time: April-July.
Plant Associates: Artemisia californica, Calystegia occidentalis, Chlorogalum pomeridianum, Mimulus aurantiacus, Nassella lepida, Salvia mellifera, Toxicodendron diversilobum.
Comments: Rare in SCR Co. A synonym of *M. fasciculatus* in *The Jepson Manual*. T238 J752

Malvaceae (mallow) ***Malacothamnus hallii***
Hall's bush mallow
CNPS List: 1B **State/Federal Status:** CEQA
Distribution: ALA?, CCA, MER, SCL
Habitat: Mostly ultramafic chaparral.
Description: Upright white-tawny shrub from 1-5 m tall. Leaf blade 2-6 cm long, thin, lower leaf surface covered with stellate scales, the epidermis clearly visible, upper surface generally sparsely hairy. Inflorescence spike-like to open panicle, many-flowered, pedicels 2-6 mm long. Calyx 4-11 mm long, densely stellate. Petals 10-15 mm long, pink-rose. Flowering time: May-July.
Plant Associates: Artemisia californica, Baccharis pilularis, Calystegia occidentalis, Ceanothus ferrisae, Centaurea melitensis, Heteromeles arbutifolia, Marrubium vulgare, Mimulus aurantiacus, Prunus ilicifolia, Salvia mellifera, Toxicodendron diversilobum.
Comments: Does plant occur in ALA Co.? A synonym of *Malacothamnus fasciculatus* in *The Jepson Manual*. T239 J752

Malvaceae (mallow) ***Sidalcea hickmanii viridis***
Marin checkerbloom
CNPS List: 1B **State/Federal Status:** /C2
Distribution: MRN, NAP, SFO, SMT, SON
Habitat: Dry ridges in coastal scrub, or ultramafic chaparral.
Description of Plant: Perennial from thick roots, stem not greater than 3 dm tall, sparsely stellate hairy. Bracts shorter than calyx, linear to narrowly oblong, generally fused at base, stipule-like; bractlets shorter than calyx. Leaves fan-shaped, generally not lobed. Flower pale pink to pink-lavender, petals up to 25 mm long. Flowering time: May-June.
Plant Associates:
Comments: Possibly threatened by development. J758

Malvaceae (mallow) ***Sidalcea malachroides***
maple-leaved checkerbloom
CNPS List: 1B **State/Federal Status:** CEQA
Distribution: HUM, MEN, MNT, SCL, SCR, OR
Habitat: Woodlands and clearings in coastal prairie, stable dunes and coniferous forest.
Description: Perennial herb or subshrub from woody root, stem from 4-15 dm tall, bristly and stellate throughout; flowers bisexual, staminate, pistillate, or mixed. Leaf blades grapeleaf-like, coarsely crenate and shallowly lobed; stipules entire. Inflorescence a much-branched panicle; units head-like; bractlets none or 1-2. Calyx 6-9 mm long, generally purplish; petals 7-15 mm long, white or purple-tinged. Flowering time: May-August.
Plant Associates: Baccharis sp., Elymus californicus, Hierochloe odorata, Salix sp., Sequoia sempervirens, Solanum douglasii.
Comments: How common is plant in HUM and MEN Counties? Endangered in OR. See *Botany of Captain Beechey's Voyage Supplement*, p. 326 (1840) for original description, and *University of Washington Publications in Biology* 18:1-97 (1957) for taxonomic treatment. T236 J758

Onagraceae (evening primrose) ***Clarkia breweri***

Brewer's clarkia

CNPS List: 4 **State/Federal Status:** CEQA?

Distribution: ALA, FRE, MER, MNT, SBT, SCL, STA

Habitat: Often ultramafic.

Description: Annual herb with decumbent or erect stems to 2 dm high. Flower hypanthium 20-35 mm long, slender, no ring of hairs within; sepals staying fused in 4's, not petal-like, green to magenta; corolla rotate, petals 1.5-2.5 cm long, pink, petal length equal to the width, outer lobes wider than the middle. Stamens 4, filaments wider above; stigma extends beyond anthers. Flowering time: April-May.

Plant Associates: Calochortus albus, Calochortus invenustus, Campanula exigua, Clarkia breweri, Claytonia gypsophiloides, Eriogonum umbellatum, Eriogonum vimineum, Fritillaria falcata, Gilia capitata, Melica sp., Phacelia sp., Pinus sabiniana, Quercus douglasii, Streptanthus breweri. Openings in woodland, chaparral and coastal scrub.

Comments: Threatened by cattle grazing, and potentially by reservoir construction. T249 J789

Onagraceae (evening primrose) ***Clarkia concinna automixa***

Santa Clara red ribbons

CNPS List: 1B **State/Federal Status:** /C2

Distribution: ALA, SCL

Habitat: Mesic shaded oak woodland.

Description: Annual herb with an erect stem, to 4 dm tall. Flower hypanthium 10-25 mm long, slender, no ring of hairs within; sepals staying fused only near tip, petal-like below, thin above, red. Corolla rotate, petals 1-2 cm long, bright pink, generally white-streaked, petal length 2 times width generally oblanceolate, 3 lobes equally wide. Stigma not beyond anthers. Flowering time: April-July.

Plant Associates: Agrostis sp., Avena barbata, Calochortus albus, Clarkia breweri, Eriogonum nudum, Eriophyllum confertiflorum, Galium sp., Mimulus aurantiacus, Phacelia imbricata, Salvia columbariae, Sedum spathulifolium, Torilis arvensis, Toxicodendron diversilobum, Trifolium willdenovii.

Comments: See *Madrono* 34(1):41-47 (1987) for original description. T249 J789

Orchidaceae (orchid) ***Cypripedium fasciculatum***
clustered lady's-slipper
CNPS List: 4 **State/Federal Status:** /C2
Distribution: BUT,DNT,HUM,NEV,PLU,SCL,SCR*,SHA,SIE,SIS,SMT,TEH,TRI,YUB,ID,OR,UT,WA+
Habitat: Open coniferous forest. Usually ultramafic seeps and streambanks.
Description: Perennial herb up to 2 dm tall with rhizomes. Leaves 2, more or less basal and opposite, up to 12 cm long. Flowers in 1-4 clusters somewhat nodding. Upper sepal greenish to brown, veins dark brown; lateral petals to 25 mm long, descending; lip yellow-green below, purple above. Flowering time: March-July.
Plant Associates: Adenocaulon bicolor, Alnus sp., Cornus nuttallii, Disporum hookeri, Polystichum munitum, Pseudotsuga menziesii, Quercus wislizenii, Smilacina racemosa, Symphoricarpos albus laevigatus, Toxicodendron diversilobum.
Comments: Many occurrences but most contain few plants. Many protected populations not reproducing. Threatened by logging and horticultural collecting. See *Proceedings of the American Academy of Arts and Sciences* 17:380 (1882) for original description, *Lindleyana* 2(1):553-57 (1987) for distributional information, and *Fremontia* 17(2):17-19 (1989) for species account. T129 J1214

Orchidaceae (orchid) ***Cypripedium montanum***
mountain lady's-slipper
CNPS List: 4 **State/Federal Status:** /C3c
Distribution: DNT,HUM,MAD,MEN,MOD,MPA,PLU,SIE,SIS,SMT,SON,TEH,TRI,TUO,OR+
Description: Perennial herb up to 7 dm tall with rhizomes. From 4 to 6 stem leaves, alternate, up to 15 cm long. Inflorescence open with 1 to 3 flowers. Upper sepal twisted or wavy, purplish, to 6 cm long; lateral petals to 6 cm long, descending; lip white, to 3 cm long. Flowering time: March-July.
Plant Associates: Lilium pardalinum, Pseudotsuga menziesii, Quercus kelloggii, Rosa californica, Smilacina racemosa, Symphoricarpos albus laevigatus.
Habitat: Moist areas in mixed-evergreen and coniferous forest.
Comments: Many protected populations on US Forest Service land not reproducing. Possibly threatened by logging. See *Fremontia* 17(2):17-19 (1989) for species account. T129 J1214

Orchidaceae (orchid) ***Piperia candida***
white-flowered rein orchid
CNPS List: 4 **State/Federal Status:** CEQA?
Distribution: DNT, HUM, MEN, SCR, SIS, SMT, SON, TRI, OR, WA+
Habitat: Open to shaded sites in coniferous and mixed evergreen forests of coastal mountains.
Description: Perennial to 5 dm tall from a bulb-like tuber. Basal lvs 7-18 cm long, 12-30 mm wide. Infl to 3 dm tall, open and 1-sided. Upper sepal pointed forward, white or green with white margins; lower white, midvein green; lateral petals pointed forward to ascending, white, midvein green; lip 1.5-3.5 mm, narrowly triangular, recurved toward spur, white; spur 1.5-3.5 mm. Flowering time: May-August.
Plant Associates: Arbutus menziesii, Lithocarpus densiflora, Piperia elongata, Piperia transversa, Sequoia sempervirens.
Comments: Difficult to identify from herbarium material. See *Lindleyana* 5(4):205-211 (1990) for original description. J1215

Orchidaceae (orchid) ***Piperia michaelii***
Michael's rein orchid
CNPS List: 4 **State/Federal Status:** CEQA?
Distribution: ALA, CCA, HUM, MNT, MRN, SBT, SCR, SCZ, SFO, SLO, SMT
Habitat: Coastal: shrub and prairie, foothill woodland, mixed-evergreen and closed-cone-pine forest.
Description: Perennial up to 7 dm tall from a bulb-like tuber. Basal leaves 7-24 cm long, 1-5 cm wide. Inflorescence dense to open, to 2 dm tall. Flowers green to yellow-green, upper sepal ascending, lower sepals spreading; lateral petals more or less ascending; lip ovate-deltate, 3-6 mm long; spur 8-12 mm long, curved and generally pointed down. Flowering time: June-August.
Plant Associates: Achillea millefolium, Artemisia californica, Aster chilensis, Baccharis pilularis, Carpobrotus sp., Castilleja sp., Daucus pusillus, Dudleya caespitosa, Erigeron glaucus, Eriogonum latifolium, Eriophyllum staechadifolium, Festuca californica, Festuca rubra, Fragaria chiloensis, Gnaphalium sp., Mimulus aurantiacus, Rubus ursinus, Toxicodendron diversilobum.
Comments: To be expected in the Sierra Nevada foothills, Western Transverse Ranges, and northern South Coast; need information. See *Bulletin of the California Academy of Sciences* 1:282 (1885) for original description, and *Botany Journal of the Linnean Society* 75:245-270 (1977) for revised nomenclature. J1216

Pinaceae (pine) ***Pinus radiata***
Monterey pine
CNPS List: 1B **State/Federal Status:** /C2
Distribution: MNT, SCR, SLO, SMT, BA, GU
Habitat: Closed-cone pine forests.
Description: Cone bearing tree. Mature bark black, deeply grooved; mature crown irregular, round-topped. Needles in bundles of 3, 6-16 cm long, dark green; sheath persistent. Seed cone recurved, 6-15 cm long, asymmetric, light brown, persisting on stems many years after maturity, opening by fire or warm weather. Seed cone scale tips, except upper, with a rounded knob less than 2 cm.
Plant Associates: Eriophyllum staechadifolium, Mimulus aurantiacus, Polypodium californicum, Quercus agrifolia, Rubus ursinus, Solanum umbelliferum, Toxicodendron diversilobum.
Comments: Only three native stands in CA. Threatened by genetic contamination, development and fragmentation, especially at Del Monte Forest (MNT Co.) and in SLO Co. Introduced stands occur in other areas of CA. Plants from BA and GU are genetically distinct. See *Fremontia* 18(2):15-21 (1990) for discussion of genetic conservation work. T63 J120

Poaceae (grass) ***Elymus californicus***
California bottle-brush grass
CNPS List: 4 **State/Federal Status:** /C3c
Distribution: MNT, MRN, SCR, SMT, SON
Habitat: Coniferous forest and moist woodland.
Description: Perennial herb up to 2 m tall. Leaf blade to 2 m wide, flat. Flowering spike up to 25 cm long, nodding. Glumes none. Awn to 2 cm long and straight. Flowering time: June-August.
Plant Associates: Galium nuttallii, Melica geyeri, Myosotis latifolia, Pseudotsuga menziesii, Rubus ursinus, Toxicodendron diversilobum, Urtica dioica gracilis.
Comments: Synonym: *Hystrix californica*. T87 J1254

Poaceae (grass) ***Hordeum intercedens***
vernal barley
CNPS List: 3 **State/Federal Status:** CEQA?
Distribution: ANA,FRE?,KNG,LAX,MNO,RIV,SBA,SBR,SBT,SCM, SCT, SCZ,SDG,SMT,SNI,SRO,VEN,BA
Habitat: Vernal pools, saline streambeds, alkaline flats and depressions.
Description: Annual grass to 4 dm tall, bent at base or erect, loosely tufted, nodes generally hairy. Leaf sheath hairs in vertical lines, appendages < 2 mm or 0; blade less than 2 mm wide, sparsely to densely long-spreading-hairy. Lemma of central spikelet with awn < 2 mm; glumes of central spikelet clearly flat near base. Inflorescence up to 6.5 cm long, pale green. Flowering time: March-June.
Comments: Location and rarity information needed. Most occurrences have been extirpated by development, others are threatened. Previously confused with *Hordeum pusillum*. See *Nordic Journal of Botany* 2:307-321 (1982) for taxonomic treatment. Synonym: *H. pusillum*. J1266

Polemoniaceae (phlox) ***Eriastrum brandegeae***
Brandegee's eriastrum
CNPS List: 1B **State/Federal Status:** /C2
Distribution: COL, GLE, LAK, SCL, TEH, TRI
Habitat: Volcanic soils in chaparral and woodland.
Description: Slender, brittle annual herb, 1-2 dm high. Upper leaves 3-cleft into linear segments; bracts 3-5 cleft; flower light blue to white, 8-9 mm long; corolla throat 1 mm long; stamens 0.7 mm long, included in the corolla throat. E. brandegeae in herbarium specimens has longer corolla throats (2 mm) and stamens (1.5 mm); but in fresh material the size differences are less clear. Flowering time: May-August.
Plant Associates: Acanthomintha lanceolata, Adenostoma fasciculatum, Aira caryophyllea, Bromus madritensis rubens, Clarkia purpurea, Eriastrum abramsii, Navarretia atractyloides, Quercus berberidifolia, Stellaria nitens, Streptanthus callistus, Trichostema sp., Vulpia bromoides, Vulpia myuros hirsuta.
Comments: Threatened by grazing, hiking trails and vehicles (ORVs). Includes *Eriastrum tracyi* which is state listed as rare. See *Madrono* 8:88-89 (1945) for original description. J826

Polemoniaceae (phlox) ***Linanthus acicularis***
bristly linanthus
CNPS List: 4 **State/Federal Status:** CEQA?
Distribution: ALA, CCA?, FRE, HUM, LAK, MEN, MRN, NAP, SMT, SON
Habitat: Chaparral and coastal prairie.
Description: Hairy annual up to 15 cm tall. Leaves 3 times divided, needle-like, 3-10 mm long. Flowers sessile in dense heads. Sepals 6-8 mm long, lobes needle-like, an inconspicuous hyaline membrane in the sinuses. Petals and floral tube yellow, to 3 cm long including throat and lobes, petal lobes 3-5 mm long. Stamens exserted. Flowering time: April-July.
Plant Associates: Aira caryophyllea, Rhamnus californica, Umbellularia californica.
Comments: Historical occurrences need verification. Does plant occur in CCA Co.? See *Pittonia* 2:259 (1892) for original description. T279 J842

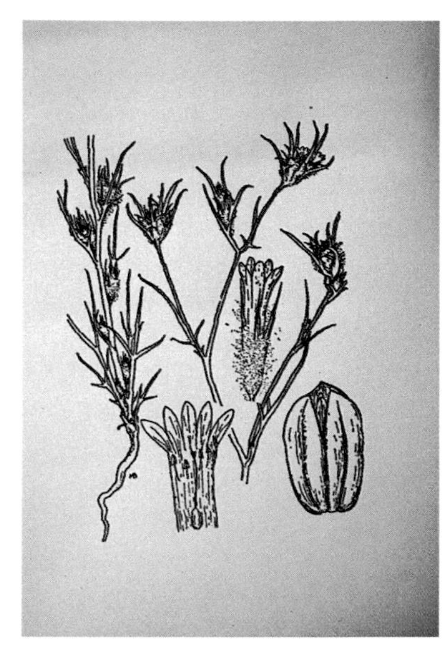

Polemoniaceae (phlox) ***Linanthus ambiguus***
serpentine linanthus
CNPS List: 4 **State/Federal Status:** CEQA?
Distribution: ALA, CCA, MER, SBT, SCL, SCR, SJQ, SMT, STA
Habitat: Mostly ultramafic grasslands, coastal scrub and foothill woodland.
Description: Annual to 2 cm, branched with thread-like stems. Leaf lobes lanceolate to 5 mm long. Flowers solitary on a peduncle to 25 mm. Sepals 3-6 mm, generally hairy with a hyaline membrane in between the sepals. Corolla funnel-shaped, tube 4-6 mm long, purple with a hairy band inside near the throat. Throat violet, lobes 4-6 mm long, lanceolate, pink with yellow base, stamens exserted. Flowering time: March-June.
Plant Associates: Brodiaea terrestris, Calystegia subacaulis, Cryptantha flaccida, Eriogonum nudum, Eriophyllum confertiflorum, Gilia clivorum, Gilia tricolor, Lomatium dasycarpum, Lotus humistratus, Lotus wrangelianus, Minuartia douglasii, Nassella lepida, Plantago erecta, Sanicula bipinnatifida, Streptanthus albidus peramoenus.
Comments: To be expected in other adjacent counties. See *Botanical Gazette* 11:339 (1886) for original description. T278 J842

Polemoniaceae (phlox) ***Linanthus grandiflorus***
large-flowered linanthus
CNPS List: 4 **State/Federal Status:** CEQA?
Distribution: ALA, KRN, MAD, MER, MNT, MRN, SBA*, SCL, SCR, SFO, SLO, SMT, SON
Habitat: Mostly on sandy soil in foothill woodland, grassland and coastal: scrub, prairie, dunes.
Description: Annual with hairy stems. Leaf lobes 1-3 cm, linear, hairy or glabrous. Flowers in dense heads, terminal, flowers sessile. Calyx 10-14 mm long, hairy, membrane wider than ribs, 2/3 length of calyx. Corolla funnel-shaped, tube 5-6 mm, white with a ring of hairs on inner surface, throat yellow, lobes 10-15 mm long, white or pink; stamens included. Flowering time: April-July.
Plant Associates: Linanthus dichotomus, Pinus sabiniana, Quercus douglasii, Quercus kelloggii, Toxicodendron diversilobum.
Comments: Many historical occurrences have been extirpated by development; need information. Other taxa often misidentified as *L. grandiflorus*. See *Pittonia* 2:260 (1892) for original description. T278 J843

Polygonaceae (buckwheat) ***Chorizanthe cuspidata cuspidata***
San Francisco Bay spineflower
CNPS List: 1B **State/Federal Status:** /C2
Distribution: ALA*, MRN, SCL?, SCR?, SFO, SMT, SON
Habitat: Sandy places in coastal: bluff, terrace, scrub, dunes, prairie.
Description: Soft-hairy, annual herb that lies down but raises up at the tip, from 5-15 cm high and to 10 dm across. Leaf blade 5-50 mm long, no stem leaves. Involucres less than 4 mm long, awns hooked near the tip. Flower 2-3 mm long, white to rose, densely hairy with a central spine at tip. Flowering time: April-July.
Plant Associates: Agoseris sp., Armeria maritima californica, Artemisia pycnocephala, Baccharis pilularis, Bromus diandrus, Camissonia cheiranthifolia, Carpobrotus sp., Erigeron glaucus, Eriogonum latifolium, Fragaria chiloensis, Lessingia germanorum, Lupinus chamissonis, Phacelia californica, Pteridium aquilinum pubescens, Rumex acetosella, Solidago sp.
Comments: Plants may occur in SCL and SCR Cos.; need more information. Closely related to *C. pungens*. See *Chorizanthe cuspidata* in *The Jepson Manual*. Some plants from Point Reyes probably intermediate to *var. villosa*. See *Proceedings of the Davenport Academy of Natural Sciences* 4:60 (1884) for original description, and *Phytologia* 66(2):98-198 (1989) for taxonomic treatment. T142 J858

Polygonaceae (buckwheat) ***Chorizanthe robusta robusta***
robust spineflower
CNPS List: 1B **State/Federal Status:** /PE
Distribution: ALA*, MNT, SCL*, SCR, SMT*
Habitat: Sandy places in coastal: scrub, dunes, strand.
Description: Soft-hairy, gray annual herb, usually erect in form. Leaf blade 1-5 cm long. Stem leaves present. Involucres over 4 mm long. Flower bracts with hooked awns, flower to 4 mm long, white to rose, hairy, lobes jagged or with a central tooth. Flowering time: May-September.
Plant Associates: Artemisia pycnocephala, Baccharis pilularis, Bromus diandrus, Dichelostemma capitatum capitatum, Ericameria ericoides, Eriophyllum staechadifolium, Quercus agrifolia.
Comments: Most populations extirpated and now known from only four extended occurrences. Threatened by development, recreation, mining, and dune stabilization. See *C. robusta* in *The Jepson Manual*. See *Phytologia* 66(2):130-131 (1989) for taxonomic treatment. T141 J859

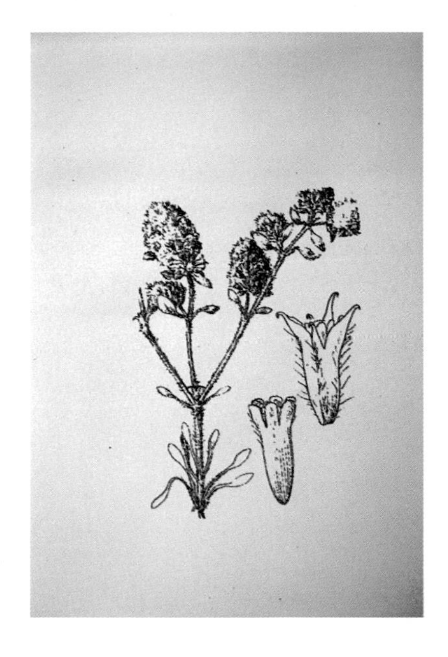

Polygonaceae (buckwheat) ***Eriogonum argillosum***
clay-loving buckwheat
CNPS List: 4 **State/Federal Status:** CEQA?
Distribution: MNT, SBT, SCL
Habitat: Bare clay or ultramafic in woodland.
Description: Annual herb 8-60 cm tall. Leaves basal with blades 5-50 mm long, usually elliptical, and tomentose (especially below). Inflorescence cyme-like, axis subglabrous; lower bracts leaf-like and scale-like, reduced upward; involucres slender-stalked (few lateral), 2-3 mm long, ribbed, glabrous, teeth 5. Flower perianth 1-2 mm long, white to rose, glabrous. Fruit 2-2.5 mm long, glabrous. Flowering time: March-June. Annual grasses, Camissonia boothii decorticans, Sisymbrium sp.
Comments: See *Phytologia* 66(4):376 (1989) for taxonomic treatment. T143 J869

Polygonaceae (buckwheat) ***Eriogonum luteolum caninum***
Tiburon buckwheat
CNPS List: 3 **State/Federal Status:** /C3c
Distribution: ALA, CCA?, COL, LAK, MRN, NAP, SCL, SMT, SON*
Habitat: Ultramafic in chaparral, valley and foothill grassland.
Description: Annual, stem up to 30 cm tall, inflorescences more or less evenly branched, or umbel-like. Leaves basal, often some on the stem. Involucres 3 mm long, teeth 5. Flowers 2 mm long, white to rose. Flowering time: July-September.
Plant Associates: Bromus carinatus, Calycadenia multiglandulosa, Chlorogalum pomeridianum, Elymus multisetus, Eschscholzia californica, Hemizonia congesta congesta, Hesperolinon congestum, Koeleria macrantha, Lomatium dasycarpum, Nassella pulchra, Phacelia imbricata, Poa secunda secunda.
Comments: Does plant occur in CCA Co.? Easily confused with *E. l. var. luteolum*; location information needed. Protected at Ring Mtn. Preserve. See *Flora Franciscana*, pp. 150-151 (1891) by E. Greene for original description, and *Phytologia* 66(4):378-379 (1989) for alternative treatment which restricts *var. caninum* to ALA and MRN Cos. Synonym: *Eriogonum vimineum var. caninum*? (Thomas). T143 J874

Portulacaceae (purslane) ***Calandrinia breweri***
Brewer's calandrinia
CNPS List: 4 **State/Federal Status:** CEQA?
Distribution: CCA,LAX,MEN,MNT,MPA,MRN,NAP,SBA,SBD,SCL, SCR, SCZ,SDG,SLO,SMT,SON,VEN,BA
Habitat: Burned or disturbed areas in chaparral and coastal scrub.
Description: Prostrate to ascending annual. Leaves to 8 cm, spoon-shaped, flat. Bracts leaf like; pedicels 6-20 cm. Sepals 4-6 mm, petals 3-5 mm, red. Mature fruit about twice as long as the sepals. Seeds shiny black, glabrous. Flowering time: March-June.
Plant Associates: Adenostoma fasciculatum, Antirrhinum cornutum, Cryptantha intermedia, Lupinus microcarpus, Nicotiana attenuata.
Comments: How common is this plant? Plant appears to be widely scattered but uncommon everywhere, and most collections are old. Field surveys needed. See *Proceedings of the American Academy of Arts and Sciences* 11:124 (1876) for original description. T158 J896

Portulacaceae (purslane) ***Calyptridium parryi hesseae***
Santa Cruz Mtns. pussypaws
CNPS List: 3 **State/Federal Status:** CEQA?
Distribution: MNT, SBT, SCL, SCR*
Habitat: Chaparral, coniferous forest and oak woodland.
Description: Annual herb from 2-11 cm high. Stems spreading to ascending. Leaves basal and cauline, 1-3 cm long. Inflorescence 1-3.5 cm long; bracts ovate to elliptic; flowers subsessile, deciduous in fruit. Sepals 2-5 mm long, ovate to round, margin with a white-scarious membrane; petals 3, 1.5-3 mm long, generally white, style 0. Flowering time: June-July.
Plant Associates: Adenostoma fasciculatum, Antirrhinum multiflorum, Arbutus menziesii, Ceanothus sp., Eriodictyon californicum, Eriogonum inerme, Pinus attenuata, Pinus ponderosa, Quercus agrifolia, Sambucus mexicana. **Comments:** Move to List 1B?Location, rarity, and endangerment information needed, especially quads for MNT and SBT counties. T160 J896

Potamogetonaceae (pondweed) ***Potamogeton filiformis***
slender-leaved pondweed
CNPS List: 2 **State/Federal Status:** CEQA
Distribution: LAS, MER, MNO, SCL*, AZ, NV, OR, ++
Habitat: Shallow, clear freshwater of lakes and drainage channels, marshes and swamps.
Description: Aquatic, slender perennial herb. Stem branched below, to 65 cm tall. Leaves all submersed, sessile to 12 cm long, to 3 mm wide, linear to thread-like; stipules up to 3 cm long, fused and sheathing leaf base, tips free, scarious. Inflorescence a spike, interrupted below, whorls 2-5, up to 2.5 cm. long; stigma sessile, not beaked. Fruit 2-2.7 mm long. Flowering time: May-July.
Plant Associates:
Comments: To be expected in the San Joaquin Valley, San Francisco Bay Area, and the central high Sierra Nevada; need information. Need quads for LAS Co. On review list in OR. J1305

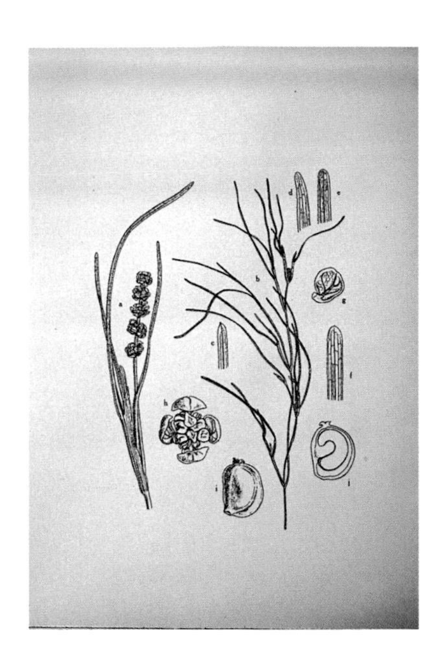
a
b
c
d
e
f
g
h
i
i

Ranunculaceae (buttercup) ***Delphinium californicum interius***
Hospital Canyon larkspur
CNPS List: 1B **State/Federal Status:** /C2
Distribution: ALA, CCA, SCL, SJQ, SLO
Habitat: Slopes in open moist woodlands.
Description: Perennial herb from 6-22 dm tall. Inflorescence generally branched, flowers generally more than 50, lowermost bracts usually leaf-like, main inflorescence axis glabrous, pedicels tips often puberulent. Sepals greenish white. Upper petals glabrous. Flowering time: April-June.
Plant Associates: Eastwoodia elegans, Forestiera pubescens, Ribes quercetorum.
Comments: See *Leaflets of Western Botany* 2:137 (1938) for original description. J918

Ranunculaceae (buttercup) ***Ranunculus lobbii***
Lobb's aquatic buttercup
CNPS List: 4 **State/Federal Status:** CEQA?
Distribution: ALA, CCA, MEN, MRN, NAP, SCL, SOL, SON
Habitat: Vernally moist areas in: oak woodland, valley and foothill grassland, and coniferous forest.
Description: Aquatic annual herb, 20-80 cm tall. Stem submersed or floating. Submersed leaf blades 8-15 mm long, generally 2-3 dissected, segments thread-like; floating or emergent leaves none or unlike. Receptacle in fruit glabrous; sepals 1.5-3 mm long; petals 4-6 mm long,1-2 mm wide, white; style 2-3 times as long as the ovaries. Fruits 2-6, body 2-3 mm long. Flowering time: March-May.
Plant Associates: Alisma plantago-aquatica, Carex sp., Epilobium ciliatum watsonii, Eryngium sp., Juncus sp., Lasthenia sp., Plagiobothrys sp., Veronica scutellata.
Comments: Threatened by urbanization and agriculture. T170 J926

Rhamnaceae (buckthorn) ***Ceanothus ferrisae***
coyote ceanothus
CNPS List: 1B **State/Federal Status:** /FE
Distribution: SCL
Habitat: Ultramafic slopes in chaparral, valley and foothill grassland.
Description: Erect evergreen shrub 1-2 m tall. Twigs puberulent. Leaves usually opposite, round to elliptical, dark green. Leaf margins short-toothed to almost entire with rounded base. Stipules at leaf base with thick, corky, persistent bases. Flowers white, sepals petaloid; petals with lower part narrowed to a long slender stalk. Seed capsule with 3 apical conspicuous horns. Flowering time: January-March.
Plant Associates: Adenostoma fasciculatum, Arctostaphylos glandulosa, Arctostaphylos glauca, Artemisia californica, Baccharis pilularis, Cirsium fontinale campylon, Eriophyllum confertiflorum, Eschscholzia californica, Garrya sp., Heteromeles arbutifolia, Lotus scoparius, Malacothamnus hallii, Marah sp., Mimulus aurantiacus, Nassella pulchra, Pinus sabiniana, Prunus ilicifolia, Quercus agrifolia, Quercus durata, Rhamnus californica, Sambucus sp., Scrophularia californica, Toxicodendron diversilobum, Zigadenus fremontii.
Comments: Known from fewer than 5 occurrences in the Mt. Hamilton Range. Threatened by expansion of Anderson Reservoir spillway, development, and changes in fire regime. See *Madrono* 2:89-90 (1933) for original description. T236 J935

Rosaceae (rose) ***Horkelia cuneata sericea***
Kellogg's horkelia
CNPS List: 1B **State/Federal Status:** /C2
Distribution: ALA*, MRN*, MNT, SBA, SCR, SFO*, SLO, SMT
Habitat: Coastal sandhills, remnant dunes, coastal scrub.
Description: Grayish green, mat forming, strong resinous scented perennial. Stem to 70 cm long. Hairs on plant dense, ascending to appressed; hairs mostly glandless or glands hidden. Lateral leaflets pinnately veined, evenly > 10 toothed. Inflorescence condensed cluster of white petaled flowers. Flowering time: April-September.
Plant Associates: Lessingia filaginifolia californica, Pteridium aquilinum pubescens, Quercus agrifolia.
Comments: Historical occurrences need field surveys. Threatened by coastal development. Occurrence on Mt. San Bruno probably last remaining location in S.F. Bay Area. Remaining plants less distinct from *H. cuneata cuneata* than those formerly occurring near San Francisco. Also may be confused with *H. marinensis*. T195 J955

Rosaceae (rose) ***Horkelia marinensis***
Point Reyes horkelia
CNPS List: 1B **State/Federal Status:** /C2
Distribution: MEN, MRN, SCR, SMT
Habitat: Sandy coastal: flats, prairie, dune and scrub.
Description: Matted perennial, grayish; odor very strong. Stem decumbent to ascending to 30 cm long. Leaves 4-10 cm long, lateral leaflets palmately veined at base, leaves unevenly few to many-toothed. Leaflets 5-10 per side, crowded, wedge-shaped, toothed and with dense hairs. Flowers in dense clusters; pedicels to 6 mm long; petals 4-6 mm long. Flowering time: May-September.
Plant Associates: Baccharis pilularis, Briza minor, Bromus hordeaceus, Chorizanthe sp., Hypochaeris radicata, Lupinus arboreus, Plantago lanceolata, Rumex acetosella, Silene californica.
Comments: Known only from fewer than twenty occurrences. Historical occurrences need field surveys. Presently may only occur from Point Reyes to Santa Cruz. Need quads for SCR and SMT Cos. See *Systematic Botany* 18(1):137-144 (1993) for distributional information. T196 J956

Rosaceae (rose) ***Potentilla hickmanii***
Hickman's cinquefoil
CNPS List: 1B **State/Federal Status:** CE/C1
Distribution: MNT, SMT, SON*
Habitat: Open pine forests in marshy areas, coastal: bluff, prairie and grassy meadows (vernally mesic).
Description: Decumbent, herbaceous perennial from 1-3 dm tall. Basal leaves pinnately compound with 4-6 pairs of leaflets to 15 mm long, palmately-cleft into 3-6 lance-like divisions. Flowers yellow in a cyme, 5 sepals and petals; slender styles > 1 mm attached just below tip of fruit. Bractlets below the sepals half as long as the sepals. Flowering time: April-August.
Plant Associates: Calamagrostis bolanderi, Carex gynodynama, Pinus radiata, Ranunculus sp., Rhynchospora globularis, Sparganium eurocarpum, Trifolium variegatum.
Comments: Recently rediscovered in San Mateo County, previously known only from two extant occurrences on the Monterey Peninsula, where seriously threatened by urbanization and recreational activities. See *Bulletin of the Torrey Botanical Club* 29:77-78 (1902) for original description, and *Fremontia* 21(1):25-29 (1993) for species account. T196 J968

Rubiaceae (madder) ***Galium andrewsii gatense***
serpentine bedstraw
CNPS List: 4 **State/Federal Status:** CEQA?
Distribution: ALA?, CCA, FRE, MNT, SBT, SCL, SLO
Habitat: Dry, rocky places on ultramafic soil in chaparral, open oak/pine woodland and coniferous forest.
Description: Perennial herb, green or silvery, matted, sometimes more or less erect, not woody above, a main stem apparent, to 22 cm tall. Leaves in whorls of 4, 4-11 mm long, more or less bristle-like, flat. Staminate inflorescence of few-flowered clusters. Pistillate inflorescence of solitary flowers in axils. Corolla rotate, yellowish. Flowering time: April-July.
Plant Associates: Arctostaphylos glauca.
Comments: Does plant occur in ALA Co? Threatened by vehicles in the Clear Creek area near San Benito Mtn. See *Brittonia* 10:186 (1958) for original description, and *Flora of California* 4(2):35-36 (1979 by W.L. Jepson for taxonomic treatment. T324 J981

Scrophulariaceae (figwort) ***Castilleja affinis neglecta***
Tiburon Indian paintbrush
CNPS List: 1B **State/Federal Status:** CT/FE
Distribution: MRN, NAP, SCL
Habitat: Open ultramafic slopes in chaparral, valley and foothill grassland.
Description: Few-branched, bristly, hemiparasitic, perennial herb from 15-60 cm tall. Leaves 20-40 mm long, +/- lanceolate; lobes 0 to 5. Inflorescence 5-30 cm long, 15-25 mm wide, generally yellow. Calyx 15-20 mm long, divided usually 1/2 in back and front, with long nonglandular and short glandular-hairs, lobes acute to rounded; corolla 18-22 mm long. Flowering time: April-June.
Plant Associates: Calamagrostis ophiditis, Ceanothus sp., Eriogonum luteolum caninum, Hemizonia congesta congesta, Hesperolinon congestum, Melica californica, Melica torreyana, Nassella pulchra, Streptanthus glandulosus pulchellus, Streptanthus niger.
Comments: Known from 6 occurrences. Protected in part at Ring Mtn. Preserve (TNC), MRN Co. Threatened by development, gravel mining, and grazing. State-listed as *C. neglecta*; USFWS also uses this name. J1018

Scrophulariaceae (figwort) ***Collinsia multicolor***
San Francisco collinsia
CNPS List: 4 **State/Federal Status:** CEQA?
Distribution: MNT, SCR, SFO, SMT
Habitat: Moist, shady woodland.
Description: Annual up to 60 cm tall, stem loosely branched, weak, sometimes trailing. Upper leaves clasping, coarsely toothed. Flowers with gland-tipped hairs, petals to 18 mm long, upper lip whitish, lower lip lavender to bluish purple. Upper pair of stamen lacking an upward-projecting awn-like process or if present less than 1 mm long. Flowering time: March-May.
Plant Associates: Aesculus californica, Dryopteris arguta, Lonicera hispidula vacillans, Pentagramma triangularis, Quercus agrifolia, Sanicula crassicaulis, Smilacina racemosa, Symphoricarpos albus laevigatus, Thalictrum fendleri polycarpum, Toxicodendron diversilobum.
Comments: Synonym: *Collinsia franciscana*. T310 J1027

Scrophulariaceae (figwort) ***Cordylanthus maritimus palustris***
Point Reyes bird's-beak
CNPS List: 1B **State/Federal Status:** /C2
Distribution: ALA*, HUM, MRN, SCL*, SMT*, SON, OR
Habitat: Coastal salt marshes.
Description: Annual herb up to 40 cm, unbranched or sparingly branched. Foliage gray-green often tinged with purple, hairy, and salt-encrusted. Leaves oblong to lanceolate to 2.5 cm long. Flower in dense spikes, each flower up to 25 mm long, upper lip white to cream, hairy, lower lip pink to purplish-red. Stamens 4, the upper filaments slender. Flowering time: June-October.
Plant Associates: Distichlis spicata, Jaumea carnosa, Limonium californicum, Salicornia virginica, Triglochin concinna.
Comments: Historical populations within the San Francisco Bay salt marshes in San Mateo and Santa Clara Counties have not been seen in many years and are presumed to no longer exist due to coastal development. Candidate for state listing in OR. See *Brittonia* 25: 135-158 (1973) for original description. T316 J1029

Scrophulariaceae (figwort) ***Pedicularis dudleyi***
Dudley's lousewort
CNPS List: 1B **State/Federal Status:** CR/C2
Distribution: MNT, SCR*, SLO, SMT
Habitat: Coniferous forest, maritime chaparral.
Description: Hairy perennial herb up to 30 cm, floppy, with above-ground portion renewed each year. Leaves pinnately compound, mostly basal, up to 26 cm long, sometimes longer than the flowering stem. Inflorescence up to 6 cm long, lower bracts longer than flowers. Corolla beaked, 2-lipped, up to 24 mm long, club-like, light pink to purple, marked darker, upper lip hooded, lower lip half or more as long as the upper lip. Flowering time: April-June.
Plant Associates: Adenocaulon bicolor, Galium sp., Hierochloe odorata, Lithocarpus densiflorus, Myrica californica, Oxalis oregana, Pseudotsuga menziesii, Rosa gymnocarpa, Scoliopus bigelovii, Sequoia sempervirens, Trientalis latifolia, Trillium ovatum, Vaccinium ovatum, Viola sempervirens.
Comments: Known from fewer than fifteen occurrences. Threatened by trampling and potential development. Plants from Arroyo de la Cruz (SLO Co.) are different and warrant further study. See *Botanical Gazette* 41:316-317 (1906) for original description. T316 J1050

Scrophulariaceae (figwort) ***Penstemon rattanii kleei***
Santa Cruz Mtns. beardtongue
CNPS List: 1B **State/Federal Status:** CEQA
Distribution: SCL, SCR
Habitat: Coniferous forests, hardwood forests and chaparral.
Description: Perennial herb, 2-12 dm tall. Main stem leaves thin, 5-14 cm long, narrowly elliptic, shallowly toothed. Inflorescence glandular. Calyx 6-7 mm long, lobes ovate. Corolla 20-30 mm long, abruptly expanded to throat, blue-violet, whitish within, glandular outside, anther sacs dehiscing full length. Flowering time: May-June.
Plant Associates: Arctostaphylos tomentosa, Chrysolepis chrysophylla, Pinus attenuata, Quercus wislizenii.
Comments: T311 J1060

Scrophulariaceae (figwort) ***Triphysaria floribunda***
San Francisco owl's-clover
CNPS List: 1B **State/Federal Status:** /C2
Distribution: MRN, SFO, SMT
Habitat: Coastal prairie, valley and foothill grasslands sometimes ultramafic.
Description: Annual plant to 30 cm tall, glabrous or sparsely stiff-hairy. Leaf up to 40 mm long, 5-9 lobed. Inflorescence generally 1-5 cm, dense; bracts 5-12 mm, 3-7 lobed. Flowers 10-14 mm, creamy white, tube slender, lower lip about equal to beak, stamen exserted. Flowering time: April-May.
Plant Associates: Allium dichlamydeum, Castilleja densiflora, Grindelia hirsutula maritima, Lasthenia macrantha, Lupinus variicolor, Sidalcea malvaeflora, Sisyrinchium bellum, Triphysaria pusilla.
Comments: Threatened by grazing and trampling. See *Systematic Botany* 16(4):644-666 (1991) for revised nomenclature. Synonym: *Orthocarpus floribunda*. J1064

Thymelaeaceae (mezereum) ***Dirca occidentalis***
western leatherwood
CNPS List: 1B **State/Federal Status:** CEQA
Distribution: ALA, CCA, MRN, SCL, SMT, SON
Habitat: Cool, moist slopes in foothill woodland and riparian forests.
Description: Deciduous shrub with pliable, smooth, leather-like bark. Leaves simple, entire, somewhat thin and bright green. Bright yellow, drooping flower appearing at the ends of the branches before the new leaves. Flowering time: January-April.
Plant Associates: Aesculus californica, Dryopteris arguta, Lonicera hispidula, Pentagramma triangularis, Quercus agrifolia, Sanicula crassicaulis, Smilacina racemosa, Symphoricarpos albus laevigatus, Thalictrum fendleri polycarpum, Toxicodendron diversilobum, Umbellularia californica.
Comments: Known from fewer than 25 occurrences. T243 J1081

RARE AND ENDANGERED PLANTS BY SCIENTIFIC NAME AND HABITAT

Scientific Name Common Name	Habitat
Acanthomintha duttonii San Mateo thorn-mint	Ultramafic grassland.
Acanthomintha lanceolata Santa Clara thorn-mint	Rocky clearings in chapparal, sometimes ultramafic.
Allium sharsmithae Sharsmith's onion	Openings in ultramafic chapparal on talus slopes.
Arabis blepharophylla coast rock cress	Rocky outcrops, steep banks in coastal: scrub and prairie.
Arctostaphylos andersonii Santa Cruz manzanita	Open sites and edges of chaparral, coniferous and evergreen forests.
Arctostaphylos imbricata San Bruno Mtn. manzanita	Chaparral, rocky slopes.
Arctostaphylos montaraensis Montara manzanita	Slopes and ridges in maritime chaparral and coastal scrub on decomposed granitic soil.
Arctostaphylos regismontana Kings Mtn. manzanita	Granite or sandstone outcrops in chaparral, coniferous and evergreen forests.
Astragalus tener tener alkali milk-vetch	Alkaline flats, vernal pools, playas, valley and foothill grassland (adobe clay).
Atriplex joaquiniana San Joaquin spearscale	Alkaline soils in chenopod scrub, valley and foothill grassland.
Azolla mexicana Mexican mosquito fern	Ponds, slow streams, wet ditches, marshes and swamps.
Balsamorhiza macrolepis macrolepis big-scale balsamroot	Valley and foothill grassland and foothill woodland slopes.
Calandrinia breweri Brewer's calandrinia	Burned or disturbed areas in chaparral and coastal scrub.
Calochortus umbellatus Oakland star-tulip	Often ultramafic in: chaparral, grassland,woodland and coniferous forests.

RARE AND ENDANGERED PLANTS BY SCIENTIFIC NAME AND HABITAT

Scientific Name Common Name	Habitat
Calyptridium parryi hesseae Santa Cruz Mtns. pussypaws	Chaparral, coniferous forest and oak woodland.
Campanula exigua chaparral harebell	Talus slopes in clearing of chaparral, often ultramafic.
Campanula sharsmithiae chaparral harebell	Ultramafic talus slopes in chaparral and foothill woodland.
Castilleja affinis neglecta Tiburon Indian paintbrush	Open ultramafic slopes in chaparral, valley and foothill grassland.
Ceanothus ferrisae coyote ceanothus	Ultramafic slopes in chaparral, valley and foothill grassland.
Chorizanthe cuspidata cuspidata San Francisco Bay spineflower	Sandy places in coastal: bluff, terrace, scrub, dunes, prairie.
Chorizanthe robusta robusta robust spineflower	Sandy places in coastal: scrub, dunes, strand.
Cirsium andrewsii Franciscan thistle	On coastal:bluffs, prairie, ravines, scrub and seeps. Sometimes ultramafic.
Cirsium fontinale campylon Mt. Hamilton thistle	Ultramafic seeps in valley and foothill grassland.
Cirsium fontinale fontinale fountain thistle	Ultramafic seeps and ravines in valley and foothill grassland.
Clarkia breweri Brewer's clarkia	Openings in woodland,chaparral and coastal scrub. Often ultramafic
Clarkia concinna automixa Santa Clara red ribbons	Mesic shaded oak woodland.
Collinsia multicolor San Francisco collinsia	Moist, shady woodland.
Cordylanthus maritimus palustris Point Reyes bird's-beak	Coastal salt marshes.
Coreopsis hamiltonii Mt. Hamilton coreopsis	Dry, exposed, rocky slopes in foothill woodland.

RARE AND ENDANGERED PLANTS BY SCIENTIFIC NAME AND HABITAT

Scientific Name Common Name	Habitat
Cupressus abramsiana Santa Cruz cypress	Sandstone or granitic cypress forest. Stands are separated by coniferous and mixed evergreen forest.
Cypripedium fasciculatum clustered lady's-slipper	Open coniferous forest. Usually ultramafic seeps and stream banks.
Cypripedium montanum mountain lady's-slipper	Moist areas in mixed evergreen and coniferous forest.
Delphinium californicum interius Hospital Canyon larkspur	Slopes in open moist woodlands.
Dirca occidentalis western leatherwood	Cool, moist slopes in foothill woodland and riparian forests.
Dudleya setchellii Santa Clara Valley dudleya	Ultramafic, rocky outcrops in valley and foothill grassland.
Elymus californicus California bottle-brush grass	Coniferous forest and moist woodland.
Equisetum palustre marsh horsetail	Freshwater marshes and swamps.
Eriastrum brandegeae Brandegee's eriastrum	Volcanic soils in chaparral and woodland.
Eriogonum argillosum clay-loving buckwheat	Bare clay or ultramafic in woodland.
Eriogonum luteolum caninum Tiburon buckwheat	Ultramafic in chaparral, valley and foothill grassland.
Eriophyllum jepsonii Jepson's woolly sunflower	Oak woodland and chaparral, sometimes ultramafic.
Eriophyllum latilobum San Mateo woolly sunflower	Ultramafic in oak woodland, exposed grassland roadcuts.
Eryngium aristulatum hooveri Hoover's button-celery	Vernal pools and lagunas.
Erysimum ammophilum coast wallflower	Coastal: dunes, strand, scrub.

RARE AND ENDANGERED PLANTS BY SCIENTIFIC NAME AND HABITAT

Scientific Name Common Name	Habitat
Erysimum franciscanum San Francisco wallflower	Ultramafic out crops and granitic cliff in grassland and coastal: dunes and scrub.
Fritillaria agrestis stinkbells	Clay depressions or ultramafic soil in chaparral, valley and foothill woodland.
Fritillaria biflora ineziana Hillsborough chocolate lily	Ultramafic grassland.
Fritillaria falcata talus fritillary	Ultramafic talus in chaparral and foothill woodland.
Fritillaria liliacea fragrant fritillary	Moist areas, often ultramafic, openhills, in valley and foothill grasslands.
Galium andrewsii gatense serpentine bedstraw	Dry, rocky places on ultramafic soil in chaparral, open oak/pine woodland and coniferous forest.
Grindelia hirsutula maritima San Francisco gumplant	Sandy or ultramafic slopes on coastal bluffs, coastal scrub and grassland.
Grindelia stricta angustifolia marsh gumplant	Tidal areas, coastal saltwater marsh.
Helianthella castanea Diablo helianthella	Open grassy sites in chaparral, coastal scrub, riparian woodland, valley and foothill grasslands.
Hemizonia parryi congdonii Congdon's tarplant	Alkaline soils in valley and foothill grassland. Sumps and disturbed sites where water collects.
Hesperolinon congestum Marin western flax	Ultramafic soils in grassland and chaparral.
Hordeum intercedens vernal barley	Vernal pools, saline streambeds, alkaline flats and depressions.
Horkelia cuneata sericea Kellogg's horkelia	Coastal sandhills, remnant dunes, coastal scrub.
Horkelia marinensis Point Reyes horkelia	Sandy coastal: flats, prairie, dune and scrub.

RARE AND ENDANGERED PLANTS BY SCIENTIFIC NAME AND HABITAT

Scientific Name Common Name	Habitat
Isocoma menziesii diabolica Satan's goldenbush	Open slopes and cliffs in foothill woodland.
Lasthenia conjugens Contra Costa goldfields	Vernal pools, moist valley and foothill grassland.
Lathyrus jepsonii jepsonii Delta tule pea	Freshwater and brackish marshes.
Legenere limosa Delta tule pea	Wet areas, vernal pools.
Lessingia arachnoidea Crystal Springs lessingia	Open ultramafic barrens, valley and foothil grasslands, coastal scrub and roadsides.
Lessingia germanorum San Francisco lessingia	Restricted to sandy soils of remnant dunes and coastal scrub.
Lessingia hololeuca woolly-headed lessingia	Ultramafic, clay soils in coastal scrub, coniferous forests, valley and foothill grasslands.
Lessingia micradenia glabrata smooth lessingia	Ultramafic soils in chaparral, valley and foothill grasslands (often roadsides).
Lilium maritimum coast lily	Bogs, gaps in closed-cone pine forest, coastal: prairie and scrub.
Limnanthes douglasii sulphurea Point Reyes meadowfoam	Meadows, freshwater marshes, vernal pools and coastal prairie.
Linanthus acicularis bristly linanthus	Chaparral and coastal prairie.
Linanthus ambiguus serpentine linanthus	Mostly ultramafic grasslands, coastal scrub and foothill woodland.
Linanthus grandiflorus large-flowered linanthus	Mostly on sandy soil in foothill woodland grassland and coastal: scrub, prairie, dunes.
Lupinus eximius San Mateo tree lupine	Coastal: bluffs,scrub.

RARE AND ENDANGERED PLANTS BY SCIENTIFIC NAME AND HABITAT

Scientific Name Common Name	Habitat
Malacothamnus arcuatus arcuate bush mallow	Ultramafic chaparral.
Malacothamnus hallii Hall's bush mallow	Mostly ultramafic chaparral.
Monardella antonina antonina San Antonio Hills monardella	Open, rocky slopes, oak woodland, chaparral.
Monardella undulata curly-leaved monardella	Inland marine sand deposits in coastal: scrub, dunes and chaparral (ponderosa pine sandhills).
Monardella villosa globosa round-headed coyote-mint	Openings in oak woodland and chaparral.
Pedicularis dudleyi Dudley's lousewort	Coniferous forest, maritime chaparral.
Penstemon rattanii kleei Santa Cruz Mtns. beardtongue	Coniferous forests, hardwood forests and chaparral.
Pentachaeta bellidiflora white-rayed pentachaeta	Ultramafic grasslands.
Perideridia gairdneri gairdneri Gairdner's yampah	Moist soil of flats,meadows, stream sides, grasslands and pine forests.
Phacelia phacelioides Mt. Diablo phacelia	Open, rocky slopes in chaparral and foothill woodland.
Pinus radiata Monterey pine	Closed-cone pine forests.
Piperia candida white-flowered rein	Open to shaded sites in coniferous and mixed evergreen forests of coastal mountains.
Piperia michaelii Michael's rein orchid	Coastal: shrub and prairie, foothill woodland, mixed-evergreen and closed-conepine forest.
Plagiobothrys chorisianus chorisianus Choris's popcorn-flower	Grassy and moist places in coastal scrub and chaparral.

RARE AND ENDANGERED PLANTS BY SCIENTIFIC NAME AND HABITAT

Scientific Name Common Name	Habitat
Plagiobothrys glaber hairless popcorn-flower	Wet, alkaline soils in valleys, coastal marshes, meadows, swamps.
Plagiobothrys myosotoides forget-me-not popcorn-flower	Chaparral.
Plagiobothrys uncinatus hooked popcorn-flower	Canyon sides, chaparral (sandy), foothill woodland, valley and foothill grassland.
Potamogeton filiformis slender-leaved pondweed	Shallow, clear freshwater of lakes and drainage channels, marshes and swamps.
Potentilla hickmanii Hickman's cinquefoil	Open pine forests in marshy areas, coastal bluff scrub and grassy meadows (vernally mesic).
Psilocarphus brevissimus multiflorus delta woolly-marbles	Vernal pools and flats.
Ranunculus lobbii Lobb's aquatic buttercup	Vernally moist areas in: oak woodland, valley and foothill grassland, and coniferous forest.
Sanicula hoffmannii Hoffmann's sanicle	Coastal scrub, coniferous forest, mixed evergreen forest; in coastal areas; often ultramafic or clay.
Sanicula saxatilis rock sanicle	Rocky ridges and talus slopes in chaparral, foothill woodland, valley and foothill grassland.
Senecio aphanactis rayless ragwort	Drying alkaline flats in foothill woodland and coastal scrub.
Sidalcea hickmanii viridis Marin checkerbloom	Dry ridges in coastal scrub, or ultramafic chaparral.
Sidalcea malachroides maple-leaved checkerbloom	Woodlands and clearings in coastal prairie, stable dunes and coniferous forest.
Silene verecunda verecunda San Francisco campion	Sand hills and rocky soils in coastal: strand, prairie, scrub.

RARE AND ENDANGERED PLANTS BY SCIENTIFIC NAME AND HABITAT

Scientific Name Common Name	Habitat
Streptanthus albidus albidus Metcalf Canyon jewel-flower	Ultramafic valley and foothill grassland.
Streptanthus albidus peramoenus most beautiful jewel-flower	Ultramafic valley and foothill grassland.
Streptanthus callistus Mt. Hamilton jewel-flower	Open chaparral and foothill woodland.
Suaeda californica California seablite	Margins of coastal salt marshes, upper intertidal marsh zone.
Trifolium amoenum showy Indian clover	Moist, heavy soils and disturbed areas in valley and foothill grassland (sometimes ultramafic).
Triphysaria floribunda San Francisco owl's-clover	Coastal prairie, valley and foothill grasslands sometimes ultramafic.
Tropidocarpum capparideum caper-fruited tropidocarpum	Alkaline soils, low hills, valley and foothill grassland.

BIBLIOGRAPHY

Abrams, L., Ferris, R.S. 1960. Illustrated Flora of the Pacific States, Washington, Oregon, and California. In Four Volumes. Stanford University Press, Stanford, CA.

Hickman, J.C., ed. 1993. The Jepson Manual: Higher Plants of California. University of California Press, Berkeley, CA.

Munz, P.A. 1968. A California Flora with Supplement. University of California Press, Berkeley, CA.

Skinner, M.W., Pavlik, B.M. eds. 1994. Inventory of Rare and Endangered Vascular Plants of California. Special Publication No. 1 (Fifth Edition). California Native Plant Society, Sacramento. CA.

Thomas, J.H. 1961. Flora of the Santa Cruz Mountains of California. A Manual of the Vascular Plants. Stanford University Press, Stanford, CA.

General Index

ABOUT THE AUTHORS

Toni Corelli, is currently the co-chair of the Rare and Endangered Plant Committee of the Santa Clara Valley Chapter of CNPS. She received her degree in Botany from San Jose State University. She is a California Native and grew up in the Bay Area and much of her botanical knowledge has come from learning and teaching about the natural history of this area.

She has her own consulting business which consists of documenting the natural resources of public open space lands in the Bay Area. She is a docent for Jasper Ridge Biological Preserve, Edgewood Natural Preserve and Pescadero Marsh. Recently she began teaching classes in "How to Use the Jepson Manual".

She leads walks and lecture programs through out the Santa Cruz Mts. for the general public, teaching about the natural resources of this beautiful area. Compiling the information for this book has taken four years and she now plans to write several local floras.

Zoe Chandik, is currently the co-chair of the Rare and Endangered Plant Committee of the Santa Clara Valley Chapter of CNPS. She received her Masters Degree from Indiana University (IU) in Education with a Botany major. Later she became the Herbarium Curator of the Deam Herbarium at IU.

In 1962 she moved to California. She worked for Dr. John H. Thomas in the Dudley Herbarium from 1972 until it moved to the California Academy of Sciences in San Francisco.

She helped the local chapter of CNPS in documenting plants for the Wunderlich lands prior to it being established as a county park. She taught Spring Wildflower Field Identification classes through the local adult education programs. She has been a CNPS member since moving to Palo Alto; and has been very active in the local chapter for the past 10+ years on numerous committees. She is planning a book on the serpentine endemics of the Bay Area.